CW00976807

Newnes 8086 Family Pocket Book

Books in the series

Newnes Computer Engineer's Pocket Book
Newnes Data Communications Pocket Book
Newnes Hard Disk Pocket Book
Newnes Microprocessor Pocket Book
Newnes MS-DOS Pocket Book

Newnes 8086 Family Pocket Book

up to and including 80486

Ian Sinclair

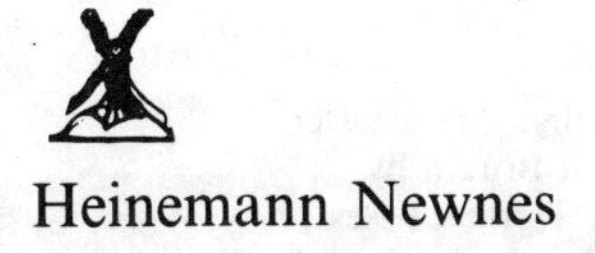

Heinemann Newnes

Heinemann Newnes
An imprint of Heinemann Professional Publishing Ltd,
Halley Court, Jordan Hill, Oxford OX2 8EJ

OXFORD LONDON MELBOURNE
AUCKLAND SINGAPORE IBADAN
NAIROBI GABORONE KINGSTON

First published 1990

British Library Cataloguing in Publication Data
Sinclair, Ian R. (Ian Robertson) *1932–*
Newnes 8086 family pocket book
1. Microprocessors
I. Title
621.3916

ISBN 0 434 91872 5

Typeset by Vision Typesetting, Manchester
Printed and bound in Great Britain by
Courier International Ltd, Tiptree, Essex

Contents

Acknowledgements

None of this book could have been written without the generous and unstinting cooperation of several leading firms, notably Intel and the distributors for Intel chips and data, Rapid Silicon Ltd. I am also very grateful to Elonex PLC for permission to reproduce the block circuit diagram of their 286M machine as an illustration of the use of the popular AT bus in computers. I am also most grateful to Matmos Ltd for the use of their excellent AT machine.

I would like also to thank my fellow authors in this series, Mike Tooley and Chris Roberts, for discussions which fixed the format of this and the other books in this series, and to all at Heinemann Newnes, particularly Peter Dixon, for their careful work on the text.

Ian Sinclair

Introduction

Note: The remarkable compatibility that exists in the Intel family of 16/32 bit processors makes it possible to cover in the space of this book the 8086, 8088, 80186, 80188, 80286, 80386 and 80486 types. The aim is to deal as fully as possible with the hardware operation, architecture and uses of these chips and their immediate support chips, but in a book of this size it is impossible to incorporate everything that is known about these chips – a full description of the 80386 chip action would fill a book of several times this size. This book, as a pocket book, exists as a portable guide to these chips rather than as an encyclopaedia, and the reader is referred to the Intel handbooks for details that might be needed by anyone designing new equipment using these chips. This is particularly important with reference to new chips for which the information in this book is taken from provisional data sheets.

Because of the upwards compatibility of the chips, descriptions of registers, buses and other features will be given in detail for the 8088/8086 types and not repeated for others, so that for the more complex chips the only details that will be examined will be for the features unique to these chips. This avoids repetition of information.

Chip history

The start of desktop computing is usually dated as 1971 when the Intel Corporation released a microprocessor chip designated as the 4004. This was a 4-bit chip and the significant feature that distinguished it from calculator chips of similar capacity was that it contained only primitive binary code arithmetical programming (microcoding) within the chip. The chip was programmable, rather than pre-programmed, so that it could be used as a universal controller rather than a specialized calculator, and it had originally been developed as part of a military contract that was cancelled, leaving Intel with a production line for manufacturing something for which there was no obvious demand.

The demand, however, rapidly appeared, and Intel developed the 4004 into the 8008, an improved version that could work with 8 bits at a time. This microprocessor had obvious applications to control circuitry, and when a more advanced version, the 8080, was developed in 1973 the way was clear for small computers to be built, though the first applications of the chips were to video games. At first, computing applications consisted of kits only which allowed enthusiasts to build their own machines. The first of these, the Altair, was advertised in January 1975. These machines were very primitive, and programming was excruciatingly awkward, with only a hex keypad for inputs and in most cases only a set of hex 7-segment displays as outputs. Nevertheless, the first of these kits achieved US sales totalling $13 000 000 in its first year, indicating a vast market for a machine even of this primitive type.

By the end of 1975, Gary Kildall, founder of Digital Research, was able to provide for computer manufacturers his operating system, a program to control, program, and monitor the action of a small computer, which was called CP/M. This system was designed to work with the 8080 chip, and machines which used other chips such as the 6502 were considerably handicapped in being unable to use CP/M because of their use of memory – CP/M required the first 100H bytes of RAM to be used for storing CP/M address vectors whereas the 6502 used this memory area as a stack. Many corporations saw the potential of the chip and the operating system, and developed machines with disk drives and using the CP/M operating system. In the years 1976 to 1980, the form of small business computers became established into a pattern that we can recognize today. The microprocessor would be of the 8-bit type, handling a maximum of 64K of memory, and the machine would use at least one 5.25 in. floppy disk drive, often two such drives and sometimes one 5.25 in. drive and one 8 in. drive.

The CP/M computers which were the standard business machines of the late 1970s suffered from the limitations of memory that were imposed by the use of the microprocessor. Contrary to popular belief, it is not the restriction to handling 8-bit bytes of data that limits these chips, but the number of bits that can be used to generate the memory reference number, called the **address** number. All of the 8-bit chips used 16 bits for address numbers, and in binary code this automatically implies the use of only 65536 addresses, 64K of memory. The use of 8-bit chips could have been considerably extended if chips had been modified to use 20 bits of address number, but by the time this could be done it was also possible to move to 16 bits of data handling, and this formed the subject of the next leap forward in desktop computing.

In 1979, Intel developed their 8086 chip, a vastly more complex microprocessor than the 8080, which could work with 16-bit units and up to 1M of memory (1M = 1024K). The 8086 was an immensely difficult chip to make successfully in those days and in consultation with IBM, who took great interest in the development, Intel designed the 8088, a version of the 8086 which read data in single bytes (8 bits) but could process it in 16-bit units. The 8088 had the advantage of being much easier to manufacture and to package, and of using standard arrangements of memory chips which were designed to work in 8-bit units.

The IBM PC was announced in November 1981, about 16 months after work had started on the design of the machine. The Intel 8088 microprocessor was used along with a set of support chips from Intel, and even by the standards of 1981 the specification of the machine was not particularly impressive compared to what could be achieved using the chips. What made the machine significant was that it could address memory up to 1M (1024K) and that it used a new operating system, PC–DOS.

Second source and workalike chips

The 8088 and 8086 chip designs were extensively licensed by Intel, and are available from a number of sources, including AMD, Fujitsu, Harris Semiconductor Corp. and Siemens. In addition, NEC have developed pin-compatible chips with identical registers but using rather different architecture and microcoding. These, V20 (replacing 8088) and V30 (replacing 8086), chips have a noticeable speed advantage as compared to the 8088 and 8086, and have a few additional instructions available, though the enhanced instructions are of little interest to anyone developing software which must maintain compatibility over the whole 16/32 bit family of Intel chips. The NEC chips are not dealt with in this book.

The 80286

Just about the time that the PC/XT machine was announced, Intel were developing the chip which would replace the 8088 and 8086 in machines that would be produced (though this might not have been foreseen) for the rest of the decade. The 8086 provided the model for the architecture which internally is almost identical to that of the 8086. The 80286, as the new chip was classified, uses 16-bit data like the 8086, but the design allows it to handle the 16-bit units much faster than the 8086. In addition, the 80286 was designed to operate at a much higher clock speed, typically 12.5 MHz. At that clock speed, Intel quoted a speed advantage of six times as compared to the 8086 at 5 MHz, and the prospect of making programs run six times faster is one that computer manufacturers cannot resist. Following Intel's policy of compatibility the new chip was constructed so that it could run any code written for the 8088 or 8086, and added only a few codes of its own.

Speed alone was not the priority, however. Even at that time, when software houses were just starting to take advantage of the memory capacity of the PC/XT machine, Intel were looking forward to the use of machines with much larger capacities, and the 80286 chip was designed to use 24 address lines, allowing it to work directly with up to 16M of memory. This, however, would not necessarily be available to existing programs. The design of the 80286 provided for the chip to be switched to work in either one of two different ways, according to the status of one bit in a register at the moment of switching it on. In one mode, called real-address mode, only 20 of the address lines would be used, and the addressing system would be identical to that of the 8086, up to 1M of total memory. In this mode, the 80286 could be used to run any program which would run on the 8086 machines, since most of its operating codes were identical. The other operating mode, called protected mode, allowed the use of up to 16M of memory, with portions of the memory separated so that several programs could be run at once without interfering with each other.

Things did not quite work out in this way, however. The snag was that the 80286 had to be switched one way or the other, either to the real-mode in which it could run standard PC–DOS or MS–DOS software, or to the protected mode in which it could not. Protected mode was of no use with PC–DOS or MS–DOS and though it could be used with other operating systems such as Unix or Xenix, the necessary operating system for the chip did not exist. Even now, no suitable operating system has been made available to take advantage of the 80286 protected mode and yet allow MS–DOS programs to be run, and such a development is unlikely now that the 80386 and 80486 chips are being used.

When IBM announced the PC/AT machine, in 1984, the advertising for the machine stressed the capabilities of the new chip and stressed that the machine could support three users, running programs under the Xenix operating system. This, however, was not what the buyers wanted. The machine was accepted eagerly, but as a way of running standard PC–DOS/MS–DOS software at high speed rather than as a multiuser machine using Xenix. Practically all of the AT machines that have been sold have spent their entire lives with the chip in its real mode, addressing 1M of memory and running the same applications programs as the XT, but faster. The additional memory that can be used, up to 16M, is accessible only as **extended** memory (not to be confused with expanded memory on the XT) and this requires suitably-written software to make use of the additional memory.

Second source chips

A more recent development has been in fast 286 chips. Intel has licensed manufacture of the 80286, as it did with the 8088 and 8086, to several other chip-makers, such as AMD, C&T, Fujitsu, Harris Semiconductor Corp. and Siemens, and both Intel and these other makers of 80286 chips have developed versions that run at much higher speeds than the original version. The first AT machines used the 80286 at a clock speed of 10 MHz, which was at the time a fairly fast speed for the chip, which was available in 6 MHz, 8 MHz, 10 MHz and 12.5 MHz versions. Since then, 16 MHz versions of the 80286 have become available and are being used by makers of AT clones. Running the 80286 at this speed makes some of these machines as fast as the slower of the 80386 machines in the main (usually the only) task that they have, which is of running MS–DOS applications software.

A few PC clone manufacturers have taken the risk of running the 80286 at 20 MHz, which is stretching its capabilities rather further than the original design intended. At this speed, a computer can retain the low-price advantages of any 286 machine, but can be very competitive in that portion of the market where speed counts. Though it seems an attractive solution to run the microprocessor chip at a high clock speed, this is by no means something that can be done easily. To start with, the power dissipation of a chip increases considerably as the clock speed is raised, so that the chip will overheat unless considerable attention is paid to cooling it. In addition, the other chips in the system may not be able to run at an increased speed, and it may be necessary for the microprocessor to wait between instructions so as to allow the memory chips to catch up. A processor that uses 'wait states' like this will run its internal processes at full speed but it will be slower at anything that involves access to memory – as most processor instructions require. Despite all this, several manufacturers have succeeded in running the 80286 at high clock speeds with zero wait states so that the resulting machines are fast and efficient. Prolonged mutterings from pundits that the 80286 chip is an evolutionary dead-end have not seriously dented the sales of machines of this type, particularly among users whose main requirement is for a faster XT rather than a revolutionary product which can offer only marginal advantages until software is rewritten (thus losing compatibility with earlier software).

The fastest AT-clones currently available use the NEAT (New Enhanced AT) chip set from C&T

Corp. This puts into a few chips all the functions that on the conventional AT type of circuit board are provided by a large mass of components. This in itself makes these actions faster, but the NEAT set will also operate with the 80286 chip running at high speeds, 16 MHz or 20 MHz according to type. In addition, the NEAT layout provides for other clock speeds to be used with other parts of the board, so that 8-bit expansion slots can be run at the 8 MHz speed that is usually the limit for 8-bit expansion cards, and memory can be run at the 10 MHz to 12.5 MHz speed that is normally the limit for all but the most expensive forms of memory. This can be done with remarkably little sacrifice in overall running speed, because the NEAT system allows memory to be divided into groups (**pages**) with data being fetched from one page while the processor is dealing with data on another page. By distributing data between memory pages, access can be made very much faster because the time interval between two successive reads from or writes to a page is more than the time that the memory requires for a read or write. The form of the chip set does not require the manufacturer of the computer to follow a rigidly set pattern of use, so that different manufacturers have made quite different uses of their machines based on the NEAT concept.

The 80386 chips

From the time that the AT was launched in 1984, development at Intel was focused on a new chip, the 80386. The aim for this chip was to avoid the limitations that had been built into the 80286, bearing in mind that customers still expected to be able to run PC–DOS/MS–DOS software. At the same time, the new chip was expected to be able to take advantage of any new operating system which could fully exploit its abilities. From the start, the 80386 chip was designed as a 32-bit chip, using 32 bits of data and also 32 bits of address number. This allowed for the use of enormous quantities of memory, 4096M, if this could be made available, plus virtual memory (managing a hard disk as if it were part of the main RAM) and it also allowed for the idea of running in different modes to be extended from the simple system that had been used in the 80286. Above all, it was possible for a computer designer to make more use of the potential of the chip than had been possible with the 80286.

The 80386 chip was a considerable step forward from the 80286, and its development required time. In addition, Intel protected the design of the chip with patents which prevented any cloning of the chip, and they did not licence manufacture to others. Unlike the 80286 chip, then, the 80386 which first emerged in 1986, has never been available from other suppliers, and this has maintained the price of the chip and of computers which use the chip, though prices have fallen considerably since the introduction of the first 80386 machines in 1987.

Given that most users of the more powerful 386 machines want still to make use of the standard MS–DOS applications programs, the most useful mode of the 80386 is its **virtual 8086** mode. This allows the memory to be partitioned into 1M sections, with a program running in each section, and some software, notably Microsoft's Windows–386 will allow a 386 machine to be used in this way. Many users, however, make use of a 386 machine simply as a faster version of the XT, and the advanced features of the 80386 chip, like those of the 80286 before it, are ignored since they cannot be used under MS–DOS.

The clock rate of 386 machines has progressed rapidly throughout the few years for which the machines have been available. At the start of the 1988, a typical clock speed for a 386 machine was 16 MHz, and this had increased to 20 MHz by the end of the year. Several 386 machines feature processor rates which could be switched, typically at 8/10/12/16/20 MHz. This was a feature that did not greatly appeal to customers, who kept the switch in its fastest position at all times. By mid-1989, a 33 MHz processor speed was announced by Acer, and several other machines operating at this speed followed. This speed probably represents the limit of the present chip, but if there should prove to be advantages perceived by users who are prepared to pay for the faster running speed and who actually need it (remembering the usual limitations that are imposed because of the slower response of memory chips) then it is possible that the clock speed could be screwed even higher.

The 80386SX

One answer to the cost of the 80386 has been provided by Intel in the form of the 80386SX chip. This takes the same route as was taken earlier when the 8088 was developed from the 8086, using an external number of bits of data that is limited compared to the internal number. Internally, the 80386SX is a 32-bit microprocessor, with all the internal facilities and operating modes of the full-blown 80386, but with only 16 external data lines. Quite apart from any other consideration, this means 16 fewer connections to be made, 16 fewer pins on the chip and 16 fewer data lines on the circuit board. Machines featuring the SX chip were rapidly put into production by Compaq, followed by a few other manufacturers. It is claimed that machines using the SX chip can be run at about 85–90 per cent of the speed of the original 80386, using the same clock speed.

The i486 (80486) chip

The i486 chip is the logical successor to the 80386 which signals an end to the dominance of the 386 for fast PCs. The chip is not simply a 386 with additions, it is a complete redesign, which allows faster computing rates to be achieved **without** using high clock rates. This does not mean that the i486 will necessarily be used at low clock rates, but that clock rates of 25 MHz will achieve much faster processing than the use of 33 MHz on the old 386 type of chip. More actions have been added, and the 486 includes its own numerical co-processing actions so that there will no longer be an empty socket next to the processor on the 486 machines. The enhancements include cache memory control, with the cache memory as part of the chip itself, further reducing the time needed for access to any data in the cache memory. In addition, the cache is divided into four sets, each holding data from different sections of the main memory. This means that the processor need only check one quarter of the total cache memory when looking for data from a given memory location. The use of cache memory makes it unnecessary for the i486 to use pipelined addressing, unlike the 80386.

Co-processors

The Intel chip set, whether it is based on the 8088, the superior 8086, the 80286 or the 80386, allows for the use of co-processors, chips which can take over the running of the computer from the main microprocessor. The only chip that is usually found as a co-processor, however, is the floating-point co-processor which for the 8088/8086 machines is the 8087. This is an expensive chip, costing in the region of £100, whose specialized action would otherwise be carried out by software with the inevitable penalty in operating speed.

The 8087 is a chip which deals with floating point actions using a combination of hardware and built-in software routines. The important point, however, is that it can work alongside the main processor, interrupting the main processor only when it needs to be fed with data or needs to pass results back into the main system. The use of the 8087 on the XT type of machine will therefore speed up working with floating-point numbers by an amazingly large factor. This is particularly true of the XT type of machine with an 8088 chip running at a low clock speed of around 4 MHz. Later 8088 clones have used faster speeds, and most of the 8086 machines use a clock speed of 8 MHz. The use of the 8087 is not quite such a marked advantage for the 8086 machines, but it still offers a very considerable improvement in speeds on these machines when programs that need floating-point computations are written to make use of the co-processor.

The addition of the 8087 co-processor chip to an XT type of machine very considerably speeds the performance of a computer which is running any program that depends heavily on floating-point arithmetic. More advanced derivatives of the 8087 are available for the machines based on the 80286 and 80386 chips, but the advantages of using these very expensive co-processors are not so obvious and clear-cut even for users of large spreadsheets. This is particularly true of the 80287, because the design on the 80286 is such that floating-point arithmetic can be carried out much more efficiently than on the older 8086 design. Another problem is that the design of the 80286 and 80287 does not allow the two processors to work together in the way that the 8086 and 8087 can work simultaneously. The result of these factors is that the 80287 chip, despite its £150 to £300 price tag (depending on clock speed) does not achieve such a dramatic increase in processing speed of spreadsheets, and for some software, albeit very few examples, may actually **reduce** the processing speed. This bizarre effect arises because **all** arithmetic computation is handed to the 80287 if it exists in its socket, and since the 80286 can handle integer arithmetic faster than the 80287, extensive use of integer arithmetic in a program can result in the 80287 processing this work more slowly than the 80286. In addition, it is usually necessary to make the 80287 run at a slower clock speed than the 80286.

The 80387 remedies some of these problems, but though its use can speed up floating-point arithmetic on a 386 machine, the increase in speed is not so dramatic as that resulting from the addition of the 8087 to the 8088/8086 machines, and many users of 386 machines would notice no really significant increase unless they used spreadsheets or CAD programs, certainly not proportionate to the cost. This does not mean that the principle of the co-processor is no longer valid, because other co-processors for fast arithmetic have been developed (such as the Weitek 1167) and are in the course of development. These promise to achieve the same order of improvements as the 8087 delivers for the 8088 machines, but only for suitably-written software and at a very high price. The 80486 chip integrated floating-point arithmetic functions into the microprocessor itself and makes the use of a co-processor unnecessary.

Notes

1 Throughout this book, the symbol # has been used to denote an active LOW input or output, replacing the older convention of using the bar above the signal name.

2 The Intel chips have used a variety of names, all of which are trademarks. Thus the 80386 is known as the i386, 386, 80386 and 80386DX; the 80386SX is known also as the 386SX, and the 80486 is known also as the i486.

3 All pin diagrams are as seen from the *top* of the chip, the chip side of the board.

1 Pin-outs, casings and DC conditions

Notes

V_{cc} means positive supply voltage, +5V, V_{ss} indicates negative supply voltage, earth or ground.

No pin diagrams are shown for the conventional DIL layouts.

Pin diagrams are shown from the top (chip side).

8088/8086 set

The Intel 16/32 bit processors have been made available in a variety of chip styles. The original 8088/8086 16-bit processors made use of a conventional 40-pin ceramic DIL package, with a plastic 40-pin DIL package available for the 8086. These processors use *N*-channel depletion-mode technology (HMOS), and the later 80C88A and 80C86A types use high-speed CMOS (CHMOS) structures with much lower power consumption, both in 40-pin plastic DIL packages.

The CMOS 80C88AL and 80C86AL types can be obtained with either the 40-pin DIL plastic package which is compatible with all the other versions, or in a 44-pin PLCC (pin-grid) type of package which is better suited to modern circuit-board layouts, including surface-mounting requirements. Both varieties of CHMOS chips are of fully static design, so that operation can continue normally at low clock speeds, and the clock can even be stopped and later resumed without corruption of any command unless the command involves the use of other chips which are not static.

The 40-pin DIL pin-out is illustrated in Figure 1.1, and the PLCC package in Figure 1.2 – note the numbering system used on the PLCC version. The pin-outs for 8088 and 8086 are compatible, but the 8088 uses only 8 pins which serve for both addressing and data, whereas the 8086 uses 16 such pins, allowing the 8086 to run a 16-line data bus as

```
Pin configuration - the # sign indicates active LOW. Minimum configuration
allocations are shown first, maximum (where applicable) second.

 1. GND          40. Vcc
 2. A14          39. A15
 3. A13          38. A16/S3
 4. A12          37. A17/S4
 5. A11          36. A18/S5
 6. A10          35. A19/S6
 7. A9           34. SSO# : High
 8. A8           33. MN/MX#
 9. AD7          32. RD#
10. AD6          31. HOLD  : RQ#/GT0#
11. AD5          30. HLDA  : RQ#/GT1#
12. AD4          29. WR#   : LOCK#
13. AD3          28. IO/M# : S2#
14. AD2          27. DT/R# : S1#
15. AD1          26. DEN#  : S0#
16. AD0          25. ALE   : QS0
17. NMI          24. INTA# : QS1
18. INTR         23. TEST#
19. CLK          22. READY
20. GND          21. RESET
```

Figure 1.1 *The pin-out for the 40-pin DIL version of the 8088/8086 chip.*

compared to the 8-line bus of the 8088. The pin designations show alternative names (enclosed in brackets) for some pins, because the 8088 and 8086 chips can be run in different modes, MAX and MIN mode, in which ten pins are used differently. The HMOS chips use a power supply of 5.0 V at typically 200 mA; the CHMOS chips have the same voltage requirement, with an operating current of about 10 mA per MHz of clock speed. Maximum dissipation for all types is 1 W.

Operating modes

Minimum mode implies that the microprocessor has command of the bus lines at all times, whereas maximum mode implies that the bus is being shared among processors. In minimum mode, the strap pin, pin 33 on the DIL package, is high and 9 other pins are used for bus control. In maximum mode, pin 33 of the DIL package is low, and an 8288 bus controller chip is used. The nine other pins are then dedicated to interfacing with the 8288 to indicate when the processor is using buses, when a bus request is pending or has been granted, for locking buses or for indicating the instructions are being queued. Computer applications of these chips always use maximum mode, and circuits using each mode are illustrated in Section 12.

In addition to implementing minimum and maximum operation, the 8088 family of chips can use data bus multiplexing options. The 8 pins which are

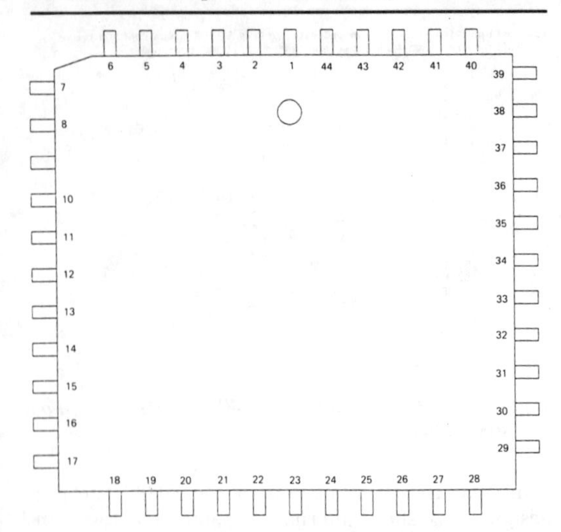

Pin configuration - the # sign indicates active LOW. Minimum configuration allocations are shown first, maximum (where applicable) second.

1. NC	44. V_{CC}
2. GND	43. A15
3. A14	42. A16/S3
4. A13	41. A17/S4
5. A12	40. A18/S5
6. A11	39. NC
7. A10	38. A19/S6
8. A9	37. SS0# ⁝ High
9. A8	36. MN/MX#
10. AD7	35. RD#
11. AD6	34. HOLD ⁝ RQ#/GT0#
12. AD5	33. HLDA ⁝ RQ#/GT1#
13. AD4	32. WR# ⁝ LOCK#
14. AD3	31. IO/M# ⁝ S2#
15. AD3	30. DT/R# ⁝ S1#
16. AD1	29. DEN# ⁝ S0#
17. AD0	28. ALE ⁝ QS0
18. NC	27. INTA# ⁝ QS1
19. NMI	26. TEST#
20. INTR	25. READY
21. CLK	24. RESET
22. GND	23. NC

Figure 1.2(a) *The 44-pin PLCC package used for some of the CHMOS chips. The view is from the top of the package. The pin assignments are shown in (b).*

used for both address and data can be connected directly to a bus (a multiplexed bus), or to a latch and transceiver circuitry which allows the use of separate address and data buses. The 8088 is more complicated in this respect, because 8 pins are used for address purposes only, another 8 for time multiplexed address and data information, and 4 for multiplexed address and status information. The 8086 is normally used with a multiplexed 16-bit address/data bus rather than with demultiplexed buses.

8088/8086 HMOS types.

Ambient temperature, chip biased...........0 °C to 70 °C
Storage temperature....................-65 °C to +150 °C
Voltage, any pin w.r.t earth...............-1 V to +7 V
Power dissipation.................................2.5 W
Max. Case temperature, plastic.....................95 °C
Max. Case temperature, CERDIP......................75 °C

80C88A/80C86A CMOS types.

Supply voltage w.r.t earth.................-0.5 to 7.0 V
Input voltage to any pin.............-0.5 to (V_{CC}+0.5) V
Output voltage from any pin.........-0.5 to (V_{CC}+0.5) V
Power dissipation.................................1.0 W
Storage temperature....................-65 °C to +150 °C
Ambient temperature, biased................0 °C to 70 °C

80C88AL/80C86AL CMOS types.

Supply voltage w.r.t earth.................-0.5 to 8.0 V
Input voltage to any pin.............-2.0 to (V_{CC}+0.5) V
Output voltage from any pin..........-0.5 to (V_{CC}+0.5) V
Power dissipation.................................1.0 W
Storage temperature....................-65 °C to +150 °C
Ambient temperature, biased................0 °C to 70 °C

Figure 1.3 *Absolute maximum (transient) ratings for 8088/8086 chips.*

8088/8086 HMOS types.

Symbol	Meaning	Min	Max	Unit	Notes
V_{IL}	Input low voltage	-0.5	+0.8	V	Max mode
V_{IH}	Input high voltage	2.0	V_{CC}+0.5	V	Min mode
V_{OL}	Output low voltage	-	0.45	V	@ 2.5 mA
V_{OH}	Output high voltage	2.4	-	V	@ -400 μA
I_{CC}	Supply current	340 - 360		mA	25 °C amb.
I_{LI}	Input leakage		±10	μA	$0 \le V_{in} \le V_{CC}$
I_{LO}	Output leakage		±10	μA	$0.45 \le V_{out} \le V_{CC}$
V_{CL}	Clock Low	-0.5	+0.6	V	
V_{CH}	Clock High	3.9	V_{CC}+1.0	V	
C_{IN}	Input buffer C		15	pF	1 MHz
C_{IO}	I/O buffer C		15	pF	1 MHz

80C88A/80C86A and 80C88AL/80C86AL CMOS types.

Symbol	Meaning	Min	Max	Unit	Notes
V_{IL}	Input low voltage	-0.5	+0.8	V	Max mode
V_{IH}	Input high voltage	2.0	V_{CC}+0.5	V	Except clock
V_{CH}	Clock high	V_{CC}-0.8	V_{CC}+0.5	V	
V_{OL}	Output low voltage	-	0.45	V	@ 2.5 mA
V_{OH}	Output high voltage	3.0	-	V	@ -2.5 mA
	"	V_{CC}-0.4	-	V	@ -100 μA
I_{CC}	Supply current	10 mA per MHz clock speed			
I_{CCS}	Standby supply current		500 μA		Unloaded
I_{LI}	Input leakage		±1.0	μA	$0 \le V_{in} \le V_{CC}$
I_{BHL}	Leakage, bus low	50	400	μA	V_{in} = 0.8 V
I_{BHH}	Leakage, bus high	-50	-400	μA	V_{in} = 3.0 V
I_{BHLO}	Overdrive, bus low		+600	μA	
I_{BHHO}	Overdrive, bus high		-600	μA	
I_{LO}	Output leakage		±10	μA	V_{out} = V_{CC} or GND
C_{IN}	Input buffer C		5	pF	1 MHz
C_{IO}	I/O buffer C		20	pF	1 MHz
C_{OUT}	Output C		15	pF	1 MHz

Note: Input buffers for all except multiplexed AD lines and RQ/GT

Figure 1.4 *Normal DC working limits for 8088/8086.*

DC conditions

The absolute maximum ratings of the 8088 and 8086 series are summarized in Figure 1.3. Note that all of these ratings apply to brief stress conditions only, and must not be encountered in normal running. All power lines should be adequately decoupled, and pull-up resistors used where needed.

The limits of normal working conditions are summarized in the table of Figure 1.4, showing for each variant the minimum and maximum limits where applicable for low and high signal voltage.

The 80186 and 80188

The 8088/8086 processors require a number of support chips for their operation, a topic that is covered in Section 7. The 80186 chip is an enhanced 8086, capable of operation at 8 MHz or 10 MHz, and incorporating into the same chip the actions of a clock generator, two independent DMA (direct memory access) channels, a programmable interrupt controller, three programmable 15-bit timers, programmable memory and peripheral chip-select logic, a programmable wait-state generator (for use with slow-access memory) and a local bus controller. By incorporating these additional functions, the need for elaborate bus structures is eliminated, along with the need to use separate MAX and MIN modes.

The 80186 was never used in computing applications, owing to the development of the 80286, but it has been extensively used as a process controller. Like all the chips in this set, it is completely code-compatible with the 8088/8086 chips, and has an additional ten new instruction types that are available for use in control applications.

The physical form of the chip is a 68-pin plastic leaded chip carrier (PLCC), Figure 1.5(a) or ceramic leadless chip carrier (LCC); or as a ceramic pin-grid array (PGA) (Figure 1.5(b)). Figure 1.6 shows the pin assignments. The chip supply requirements are 5.0 V at typically 600 mA.

There is also a CHMOS version of the 80186, the 80C186, available in 10 MHz, 12.5 MHz and 16 MHz versions. The 80188 is a version of the 80186 with an

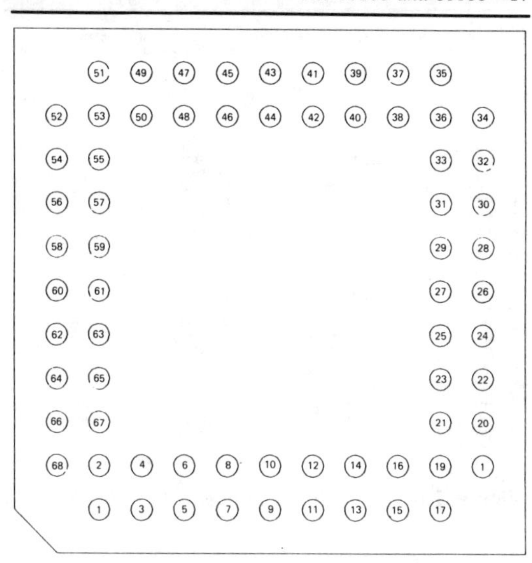

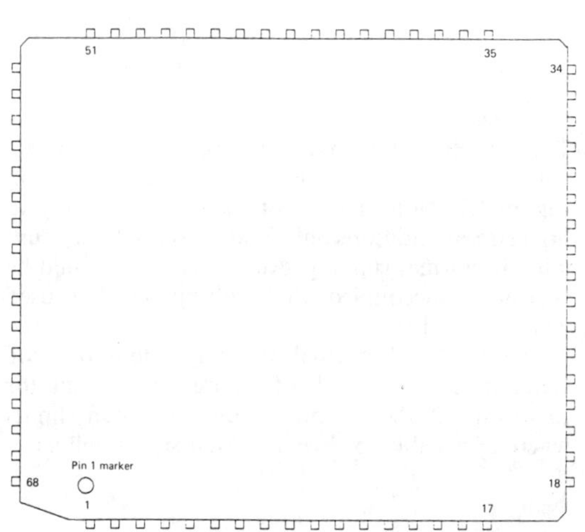

Figure 1.5 *The form of LCC, PLCC (a) and PGA (b) packages for the 80188/80186.*

8-bit data bus capability, and there is a CHMOS version, the 80C188. These versions also permit enhanced actions, see later.

Pinout for 80186 is illustrated here. The 80188 differs in using A8 - A15 in place of AD8 - AD15, and pin 64 is S7, not BHE. The # symbol indicates active LOW.

1. AD15	24. RES#	47. TEST#
2. AD7	25. PCS0#	48. LOCK#
3. AD14	26. V_{SS}	49. SRDY
4. AD6	27. PCS1#	50. HOLD
5. AD13	28. PCS2#	51. HLDA
6. AD5	29. PCS3#	52. S0#
7. AD12	30. PCS4#	53. S1#
8. AD4	31. PCS5#/A1	54. S2#
9. V_{CC}	32. PCS6#/A2	55. ARDY
10. AD11	33. LCS#	56. CLKOUT
11. AD3	34. UCS#	57. RESET
12. AD10	35. MCS3#	58. X2
13. AD2	36. MCS2#	59. X1
14. AD9	37. MCS1#	60. V_{SS}
15. AD1	38. MCS0#	61. ALE/QS0
16. AD8	39. DEN#	62. RD#/QSMD#
17. AD0	40. DT/R#	63. WR#/QS1
18. DRQ0	41. INT3/INTA1#	64. BHE#
19. DRQ1	42. INT2/INTA0#	65. A19/S6
20. TMR IN 0	43. V_{CC}	66. A18/S5
21. TMR IN 1	44. INT 1	67. A17/S4
22. TMR OUT 0	45. INT 0	68. A16/S3
23. TMR OUT 1	46. NMI	

Figure 1.6 *The pin assignment for 80188/80186 chips.*

DC conditions

The absolute maximum ratings of the 80186, 80C186, 80188 and 80C188 are summarized in Figure 1.7. Note that all of these ratings apply to brief stress conditions only, and must not be encountered in normal running. All power lines should be adequately decoupled, and pull-up resistors used where needed.

The limits of normal working conditions are summarized in the table of Figure 1.8, showing for each variant the minimum and maximum limits where applicable for low and high signal voltage.

Ambient temperature, chip biased	0 °C to 70 °C
Storage temperature	-65 °C to +150 °C
Pin voltage range w.r.t earth	-1.0 V to +7 V
Power dissipation	3 W

Figure 1.7 *The absolute maximum DC ratings for 80188/80186 chips in HMOS and CHMOS forms.*

80188/80186 HMOS and 80C188/80C186 CMOS types

Symbol	Meaning	Min	Max	Unit	Notes
V_{IL}	Input low voltage	-0.5	+0.8	V	
V_{IH}	Input high voltage (except X1 and RES#)	2.0	V_{CC}+0.5	V	
V_{IH1}	IH for X1 and RES#	3.0	V_{CC}+0.5		
V_{OL}	Output low voltage	-	0.45	V	@ 2.0 mA
V_{OH}	Output high voltage	2.4	-	V	@ -400 μA
I_{CC}	Supply current	415 -	600	mA	+70 °C to -40 °C
I_{LI}	Input leakage		±10	μA	$0 \leq V_{in} \leq V_{CC}$
I_{LO}	Output leakage		±10	μA	$0.45 \leq V_{out} \leq V_{CC}$
V_{CLO}	Clock out low		+0.6	V	
V_{CHO}	Clock out high	4.0		V	
V_{CLI}	Clock input low	-0.5	0.6	V	
V_{CHI}	Clock input high	3.9	V_{CC}+1.0	V	
C_{IN}	Input C		10	pF	1 MHz
C_{IO}	I/O C		20	pF	1 MHz

Figure 1.8 *The normal working limits for the 80188/80186 chips.*

80286 chip

The 80286 chips are made using CHMOS technology and are available in three packages, all of which use a 68-pin arrangement. The packages are the leadless chip carrier (LCC), the pin–grid array (PGA) and the plastic leaded chip carrier (PLCC) types. These packages are identical to those used for the 80186/80188 (see Figure 1.5(a) and 1.5(b)). The pin assignment is illustrated in Figure 1.9. Power requirements are for 5.0 V with typically 600 mA operating current.

The pin assignment table uses the hash (#) to indicate active LOW.

1. BHE#
2. NC
3. NC
4. S1#
5. S0#
6. PEACK#
7. A23
8. A22
9. V_{SS}
10. A21
11. A20
12. A19
13. A18
14. A17
15. A16
16. A15
17. A14
18. A13
19. A12
20. A11
21. A10
22. A9
23. A8
24. A7
25. A6
26. A5
27. A4
28. A3
29. RESET
30. V_{CC}
31. CLK
32. A2
33. A1
34. A0
35. V_{SS}
36. D0
37. D8
38. D1
39. D9
40. D2
41. D10
42. D3
43. D11
44. D4
45. D12
46. D5
47. D13
48. D6
49. D14
50. D7
51. D15
52. CAP
53. ERROR#
54. BUSY#
55. NC
56. NC
57. INTR
58. NC
59. NMI
60. V_{SS}
61. PEREQ
62. V_{CC}
63. READY#
64. HOLD
65. HLDA
66. COD/INTA#
67. M/IO#
68. LOCK#

Figure 1.9 *The pin assignment for the 80286 PLCC package, whose pin arrangement is the same as for the 80188/80186 (Figure 1.5).*

Unlike the 8088/8086 processors, no dual use of pins is necessary to operate the 80286 either as a sole processor or as part of a multiprocessor system, because the pin allocation is sufficient to allow dedicated pins to be used for these purposes. There is no pin used for switching between real mode (emulating the action of the 8088/8086) and protected mode (using the full set of address pins to address 16 Mb of memory and up to 4 Gb of virtual memory. The switching is achieved by using one bit of a word, the machine status word, stored in memory. The RESET or initial switch-on of the chip enforces the use of real mode, and once protected mode has been started by altering the machine status word, only a RESET will clear the protection bit.

In real mode, the 80286 behaves like the 8088/8086, with the top four address pins unused. The instruction set is a superset of the 8088/8086 set, so that any program that runs on the earlier chips will run on the 80286. Protected mode uses all 24 address lines, but since programs for the 8088/8086 use MS–DOS which assumes the use of the 20-bit addressing system of the earlier chips, this extended memory cannot be used in any straightforward way. The later operating system OS/2 gets around the problems by altering the machine status word at a critically-timed point in a reset cycle, so that protected mode is being used but with the ability to work as if real mode were being used.

DC conditions

The absolute maximum ratings of the 80286 are summarized in Figure 1.10. Note that all of these ratings apply to brief stress conditions only, and must not be encountered in normal running. All power lines should be adequately decoupled, and pull–up resistors used where needed.

The limits of normal working conditions are summarized in the table of Figure 1.11, showing for

```
Ambient temperature, chip biased...............0 °C to +70 °C
Storage temperature.........................-65 °C to + 150 °C
Pin voltage limits..........................-1.0 V to +7.0 V
Power dissipation......................................3.3 W
```

Figure 1.10 *Absolute maximum DC ratings for the 80286.*

Conditions: V_{CC} = 5.0 V, T_{CASE} = 0 °C to +85 °C at min. operating frequency.

Symbol	Meaning	Min	Max	Unit	Notes
V_{IL}	Input low voltage	-0.5	+0.8	V	
V_{IH}	Input high voltage	2.0	V_{CC}+0.5	V	
V_{ILC}	CLK input low	-0.5	+0.6	V	
V_{IHC}	CLK input high	3.8	V_{CC}+0.5	V	
V_{OL}	Output low voltage	-	0.45	V	@ 2.0 mA
V_{OH}	Output High voltage	2.4	-	V	@ -400 μA
I_{LI}	Input leakage		±10	μA	$0 \leq V_{in} \leq V_{CC}$
I_{LO}	Output leakage		±10	μA	$0.45 \leq V_{out} \leq V_{CC}$
I_{IL}	$I_{sustain}$ for BUSY and ERROR	30	500	μA	V_{in} = 0 V

Figure 1.11 *The normal working limits for the 80286.*

each variant the minimum and maximum limits where applicable for low and high signal voltage.

The 80386 (386, 80386DX, i386)

The 80386, like the 80286, is made using CHMOS technology, and is a full 32-bit processor, though as used in computing it runs mainly 16-bit software written for the 8088/8086 chips. The full potential of the chip is not really used in most computing uses of the 386, as the 80386 is now designated by Intel as a registered trade mark.

The mechanical form of the chip is as a 132-pin pin–grid array, and Figure 1.12 shows the physical arrangement along with the letter–number method of identifying pins. Suitable low insertion force sockets are recommended. The top surface of the package incorporates a metal panel for heat sink use.

Given the large number of pins, no multiplexing is needed, and the full 32 address lines and 32 data lines are implemented. Computer applications generally treat the 386 chip as a form of 80286, ignoring the top 16 data lines until 32-bit software is available. The processor signals do not occupy all 132 of the pins, and a large number, 42 in all, consist of supply connections with another 8 pins unconnected. **All** of the supply pins must be connected, 20 to V_{cc} and 21 to V_{ss}. No connection must ever be made to pins labelled as NC, since these may have internal connections which must not be externally connected. Future versions of the chip may use one or more of these pins for signal applications.

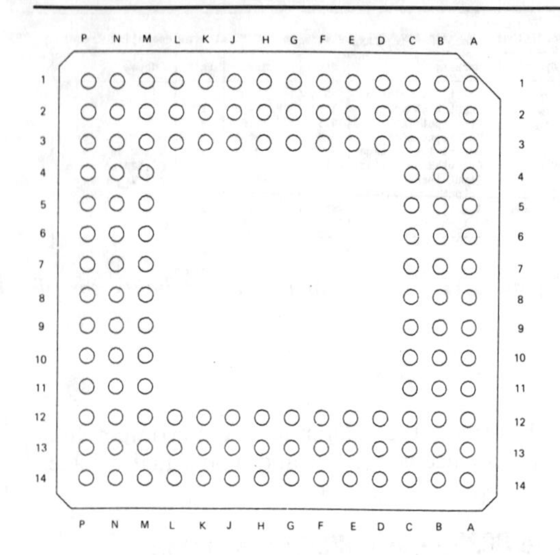

The hashmark (#) is used to identify pins whose action is active LOW.

Pin	Signal	Pin	Signal	Pin	Signal	Pin	Signal
N2	A31	M5	D31	A1	V_{CC}	A2	V_{SS}
P1	A30	P3	D30	A5	V_{CC}	A6	V_{SS}
M2	A29	P4	D29	A7	V_{CC}	A9	V_{SS}
L3	A28	M6	D28	A10	V_{CC}	B1	V_{SS}
N1	A27	N5	D27	A14	V_{CC}	B5	V_{SS}
M1	A26	P5	D26	C5	V_{CC}	B11	V_{SS}
K3	A25	N6	D25	C12	V_{CC}	B14	V_{SS}
L2	A24	P7	D24	D12	V_{CC}	C11	V_{SS}
L1	A23	N8	D23	G2	V_{CC}	F2	V_{SS}
K2	A22	P9	D22	G3	V_{CC}	F3	V_{SS}
K1	A21	N9	D21	G12	V_{CC}	F14	V_{SS}
J1	A20	M9	D20	G14	V_{CC}	J2	V_{SS}
H3	A19	P10	D19	L12	V_{CC}	J3	V_{SS}
H2	A18	P11	D18	M3	V_{CC}	J12	V_{SS}
H1	A17	N10	D17	M7	V_{CC}	J13	V_{SS}
G1	A16	N11	D16	M13	V_{CC}	M4	V_{SS}
F1	A15	M11	D15	N4	V_{CC}	M8	V_{SS}
E1	A14	P12	D14	N7	V_{CC}	M10	V_{SS}
E2	A13	P13	D13	P2	V_{CC}	N3	V_{SS}
E3	A12	N12	D12	P8	V_{CC}	P6	V_{SS}
D1	A11	N13	D11			P14	V_{SS}
D2	A10	M12	D10				
D3	A9	N14	D9	F12	CLK2	A4	NC
C1	A8	L13	D8			B4	NC
C2	A7	K12	D7	E14	ADS#	B6	NC
C3	A6	L14	D6			B12	NC
B2	A5	K13	D5	B10	W/R#	C6	NC
B3	A4	K14	D4	A11	D/C#	C7	NC
A3	A3	J14	D3	A12	M/IO#	E13	NC
C4	A2	H14	D2	C10	LOCK#	F13	NC
A13	BE3#	H13	D1				
B13	BE2#	H12	D0	D13	NA#	C8	PEREQ
C13	BE1#			C14	BS16#	B9	BUSY#
E12	BE0#			G13	READY#	A8	ERROR#
		D14	HOLD				
C9	RESET	M14	HLDA	B7	INTR	B8	NMI

Figure 1.12 *The form (a) of the pin-grid array for the 80386DX, with numbering system. The pin assignment is shown in (b).*

The thermal resistance from junction to case is 2°C/W and the casing to ambient thermal resistance can range from 19°C/W with no heat sink to 15°C/W with an unidirectional heat sink and no external cooling. When fan cooling can be used, these latter figures can be considerably reduced, with the figure dropping to as low as 5°C/W for an unidirectional heat sink in an airflow of 4 m/s velocity. Like all CMOS devices, the dissipation of the 386 depends on the operating frequency, but at least 2.5 W should be allowed for.

The 386 can use two main modes of operation, real mode and protected mode. In real mode the chip behaves as a very fast 8086 with the optional use of 32-bit internal registers, and this is the way that the chip is most frequently used in computers. The protected mode can use other operating systems that permit the use of full 32-bit software, or it can be used in a submode called virtual 8086 mode. In this submode the chip can assign memory in multiple 1 Mb sections, each of which can run an 8086 program, and the chip will use its protection mechanisms to ensure that these different tasks do not interfere with each other. The capability of the chip to use virtual memory in protected mode allows the use of up to 64 Tb of memory, where 1 Tb (terabyte) is 2^{40} or about 10^{12}. the real-memory addressing capability amounts to 2^{32} which is 4 Gb.

DC conditions

The absolute maximum ratings of the 80386 are summarized in Figure 1.13. Note that all of these ratings apply to brief stress conditions only, and must not be encountered in normal running. All power lines should be adequately decoupled, and pull-up resistors used where needed.

The limits of normal working conditions are summarized in the table of Figure 1.14, showing for each variant the minimum and maximum limits where applicable for low and high signal voltage.

```
Case temperature, chip biased..................-65 °C to +110 °C
Storage temperature............................-65 °C to + 150 °C
Supply voltage limits w.r.t Vss.................-0.5 V to +6.5 V
 Voltages on other pins...................-0.5 V to ( Vcc+0.5) V
```

Figure 1.13 *Absolute maximum ratings for the 80386DX chip.*

Conditions: V_{CC} = 5.0 V ±5%, T_{CASE} = 0 °C to +85 °C

Symbol	Meaning	Min	Max	Unit	Notes
V_{IL}	Input Low voltage	-0.3	+0.8	V	a
V_{IH}	Input High voltage	2.0	V_{CC}+0.3	V	
V_{ILC}	CLK2 input Low	- 0.3	+0. 8	V	a
V_{IHC}	CLK input High				
	- 16 and 20 MHz	V_{CC}-0.8	V_{CC}+0.3	V	
	- 25 Mhz	3.7	V_{CC}+0.3	V	
V_{OL}	Output Low voltage	-	0.45	V	@ 2.0 mA
V_{OH}	Output High voltage	2.4	-	V	@ -400 μA
I_{LI}	Input leakage		±15	μA	b
I_{IH}	Input leakage PEREQ		200	μA	
I_{IL}	Input leakage for BS16#, BUSY#, ERROR#		-400	μA	
I_{LO}	Output leakage		±15	μA	$0.45 \le V_{out} \le V_{CC}$
I_{CC}	Supply current:				
	CLK2 = 32 MHz		450	mA	Typ. 370 mA
	CLK2 = 40 MHz		500	mA	Typ. 460 mA
	CLK2 = 50 MHz		550	mA	Typ. 480 mA
C_{IN}	Input capacitance		10	pF	not 100% tested
C_{OUT}	Output or I/O capacitance		12	pF	not 100% tested
C_{CLK}	CLK2 capacitance		20	pF	not 100% tested

Notes:
a The minimum value is not 100% tested.
b V_{in} between V_{CC} and 0 V, all pins except BS16#, PEREQ, BUSY# and ERROR#

Figure 1.14 *Limits of normal working conditions for the 80386DX chip.*

80386SX

The chip which was formerly designated as 80386P9 is now referred to as the 80386SX. It uses internal 32-bit architecture and processing, but with a 16-bit external data bus and 24-bit external address bus, so that its relationship to the 386 is rather similar to the relationship of the 8088 to the 8086. The chip packaging is designated as a 100-pin fine-pitch quad flatpack (Figure 1.15), which uses 14 V_{cc} pins, 18 V_{ss} pins and 11 unconnected pins. The unconnected pins should *not* be connected to earth or to V_{cc} since they can be used for intermediate internal connections. Future versions of the chip may use one or more of these pins for external connections.

The thermal resistance, junction to case, is quoted as 7°C/W and with no heat sinking or fan assistance, the chip is rated for a maximum ambient working temperature of 33°C. This means that most applications will need either heat sinking, fan cooling or both. Without a heat sink, an airflow rate of 5 m/s is needed to permit operation in a maximum ambient of 65°C.

DC conditions

The absolute maximum ratings of the 80386SX are as for the 80386DX (see Figure 1.13). Note that all of

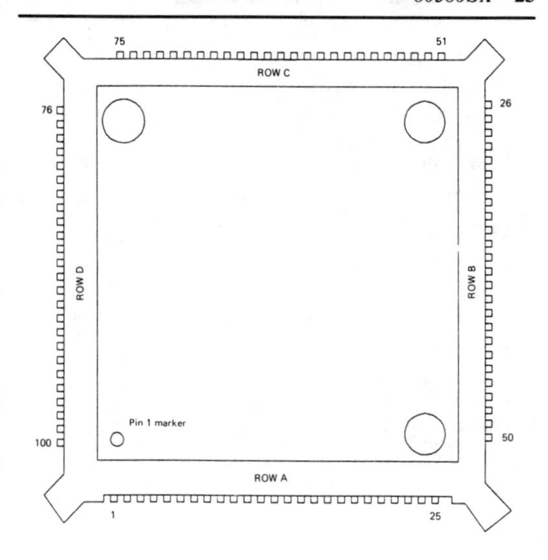

Row A Pin	Row A Signal	Row B Pin	Row B Signal	Row C Pin	Row C Signal	Row D Pin	Row D Signal
1	D0	26	LOCK#	51	A2	76	A21
2	V_{SS}	27	NC	52	A3	77	V_{SS}
3	HLDA	28	NC	53	A4	78	V_{SS}
4	HOLD	29	NC	54	A5	79	A22
5	V_{SS}	30	NC	55	A6	80	A23
6	NA#	31	NC	56	A7	81	D15
7	READY#	32	V_{CC}	57	V_{CC}	82	D14
8	V_{CC}	33	RESET	58	A8	83	D13
9	V_{CC}	34	BUSY#	59	A9	84	V_{CC}
10	V_{CC}	35	V_{SS}	60	A10	85	V_{SS}
11	V_{SS}	36	ERROR#	61	A11	86	D12
12	V_{SS}	37	PEREQ	62	A12	87	D11
13	V_{SS}	38	NMI	63	V_{SS}	88	D10
14	V_{SS}	39	V_{CC}	64	A13	89	D9
15	CLK2	40	INTR	65	A14	90	D8
16	ADS#	41	V_{SS}	66	A15	91	V_{CC}
17	BLE#	42	V_{CC}	67	V_{SS}	92	D7
18	A1	43	NC	68	V_{SS}	93	D6
19	BHE#	44	NC	69	V_{CC}	94	D5
20	NC	45	NC	70	A16	95	D4
21	V_{CC}	46	NC	71	V_{CC}	96	D3
22	V_{SS}	47	NC	72	A17	97	V_{CC}
23	M/IO#	48	V_{CC}	73	A18	98	V_{SS}
24	D/C#	49	V_{SS}	74	A19	99	D2
25	W/R#	50	V_{SS}	75	A20	100	D1

Figure 1.15 *The 100-pin fine-pitch quad flatpack for the 80386SX chip (a), and its pin assignment (b).*

these ratings apply to brief stress conditions only, and must not be encountered in normal running. All power lines should be adequately decoupled, and pull-up resistors used where needed.

The limits of normal working conditions are also as for the 80386DX (see Figure 1.14).

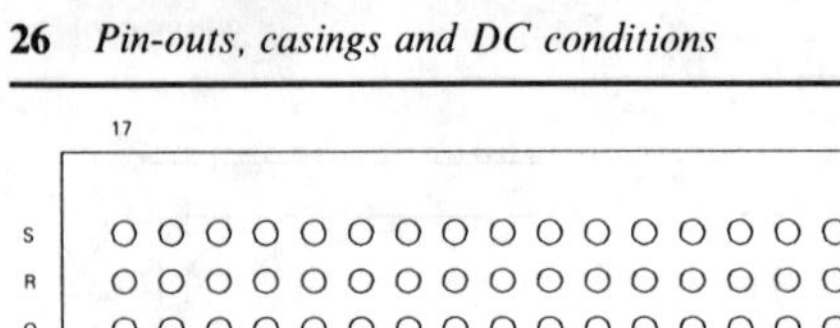

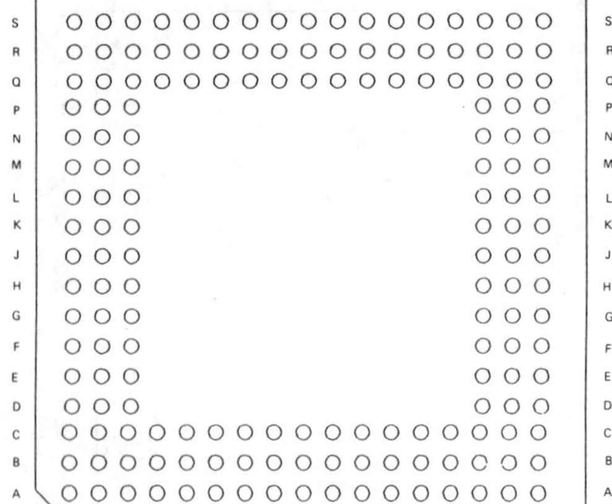

Pin	Signal	Pin	Signal	Pin	Signal	Pin	Signal
Q14	A2	C17	BS16#	P15	HLDA	B4	V_{SS}
R15	A3	C3	CLK	E15	HOLD	B5	V_{SS}
S16	A4	P1	D0	A15	IGNNE#	E1	V_{SS}
Q12	A5	N2	D1	A16	INTR	E17	V_{SS}
S15	A6	N1	D2	F15	KEN#	G1	V_{SS}
Q13	A7	H2	D3	N15	LOCK#	G17	V_{SS}
R13	A8	M3	D4	N16	M/IO#	H1	V_{SS}
Q11	A9	J2	D5	B15	NMI	H17	V_{SS}
S13	A10	L2	D6	J17	PCD	K1	V_{SS}
R12	A11	L3	D7	Q17	PCHK#	K17	V_{SS}
S7	A12	F2	D8	L15	PWT	L1	V_{SS}
Q10	A13	D1	D9	Q16	PLOCK#	M1	V_{SS}
S5	A14	E3	D10	F16	RDY#	M17	V_{SS}
R7	A15	C1	D11	C16	RESET	P17	V_{SS}
Q9	A16	G3	D12	B7	V_{CC}	Q2	V_{SS}
Q3	A17	D2	D13	B9	V_{CC}	R4	V_{SS}
R5	A18	K3	D14	B11	V_{CC}	S6	V_{SS}
Q4	A19	F3	D15	C4	V_{CC}	S8	V_{SS}
Q8	A20	J3	D16	C5	V_{CC}	S9	V_{SS}
Q5	A21	D3	D17	E2	V_{CC}	S10	V_{SS}
Q7	A22	C2	D18	E16	V_{CC}	S11	V_{SS}
S3	A23	B1	D19	G2	V_{CC}	S12	V_{SS}
Q6	A24	A1	D20	G16	V_{CC}	S14	V_{SS}
R2	A25	B2	D21	H16	V_{CC}	N17	W/R#
S2	A26	A2	D22	J1	V_{CC}		
S1	A27	A4	D23	K2	V_{CC}	A3	NC
R1	A28	A6	D24	K16	V_{CC}	A10	NC
P2	A29	B6	D25	L16	V_{CC}	A12	NC
P3	A30	C7	D26	M2	V_{CC}	A13	NC
Q1	A31	C6	D27	M16	V_{CC}	A14	NC
D15	A20M#	C8	D28	P16	V_{CC}	B10	NC
S17	ADS#	A8	D29	R3	V_{CC}	B12	NC
A17	AHOLD	C9	D30	R6	V_{CC}	B13	NC
K15	BE0#	B8	D31	R8	V_{CC}	B14	NC
J16	BE1#	M15	D/C#	R9	V_{CC}	B16	NC
J15	BE2#	N3	DP0	R10	V_{CC}	C10	NC
F17	BE3#	F1	DP1	R11	V_{CC}	C11	NC
R16	BLAST#	H3	DP2	R14	V_{CC}	C12	NC
D17	BOFF#	A5	DP3	A7	V_{SS}	C13	NC
H15	BRDY#	B17	EADS#	A9	V_{SS}	G15	NC
Q15	BREQ	C14	FERR#	A11	V_{SS}	R17	NC
D16	BS8#	C15	FLUSH#	B3	V_{SS}	S4	NC

Figure 1.16 *The 168-lead PGA package (a) for the i486, along with the pin assignment list (b).*

i486 (80486)

The replacement for the 80386DX is the i486, which is a development of the 80386DX design which incorporates built-in cache memory (and cache management) and also a built-in floating-point numeric unit. This allows the chip-count of a machine using the i486 to be reduced considerably as compared to a 386 machine of comparable performance. The performance is stated to be twice that of the 80386DX running comparable software. In addition to the much greater scale of integration, the timings for the most frequently used instructions have been improved.

The i486 uses a 168-lead ceramic pin-grid array package which is illustrated in Figure 1.16. As for the 386 chips, all of the V_{cc} and V_{ss} pins must be appropriately connected, and pins which are labelled as NC must not be connected. The quoted power dissipation at a clock speed of 25 MHz is 5 W, and at 33 MHz is 6 W. Tentative DC ratings are shown in Figure 1.17, but reference should be made to Intel for most recent information before designing around the i486.

```
Case temperature, chip biased..........0 °C to 85 °C
Storage temperature.................-65 °C to +150 °C
Voltage, any pin w.r.t earth......-0.5 V to Vcc+0.5 V
Power dissipation, 25 MHz.........................5 W
Power dissipation, 33 MHz.........................6 W
```

Figure 1.17 *Tentative absolute maximum DC ratings for the i486.*

Symbol	Meaning	Min	Max	Unit	Notes
V_{IL}	Input low voltage	-0.3	+0.8	V	
V_{IH}	Input high voltage	2.0	V_{CC}+0.3	V	
V_{OL}	Output low voltage	-	0.45	V	1
V_{OH}	Output high voltage	2.4	-	V	2
I_{CC}	Supply current		900	mA	25 MHz
			1200	mA	33 MHz
I_{LI}	Input leakage		±15	μA	3
I_{IH}	Input leakage		200	μA	4
I_{IL}	Input leakage		-400	μA	5
I_{LO}	Output leakage		±15	μA	
C_{IN}	Input C		16	pF	
C_O	I/O, output C		16	pF	
C_{CLK}	CLK C		16	pF	

Figure 1.18 *Limits of normal working for the i486 (tentative).*

2 Internal architecture – block diagrams

Note: Unless otherwise stated, all bus lines of all chips are three-state and will float to the OFF third state when not in use.

A 8088/8086 series

Though the 8088 and 8086 are pin-compatible and code-compatible, there are significant differences in the internal architecture that make the 8086 a faster processor, and the later processors are based on the 8086 architecture rather than on that of the 8088. The block diagrams that Intel supply make the differences seem greater because of the use of different drawing conventions. The compatibility is sufficient to allow a description of the 8088 to serve for both chips, with only a few notes on the differences that affect the 8086.

The 8088

The block diagram for the 8088 chips is illustrated in Figure 2.1. The main internal data bus will, because of the use of only 8 data pins, handle only 8 bits at a time, though all of the registers are 16-bits wide. The chip can be considered in two main parts, the data handling section (execution unit) and the addressing section (bus interface unit or BIU). In the data handling section, the execution register unit is comprised of the registers (see Section 3) which are used in data processing, and the arithmetic and logical unit (ALU) deals with those instructions which involve arithmetic, shifting and rotation, bit comparison and similar actions. The flag register is shown here as an extension of the ALU.

On the addressing section, the addressing registers are shown as one block, all of which are 16-bit registers. The formation of a 20-bit address is dealt with by an adder which is not specifically shown in this diagram. This forms an address by using a 16-bit

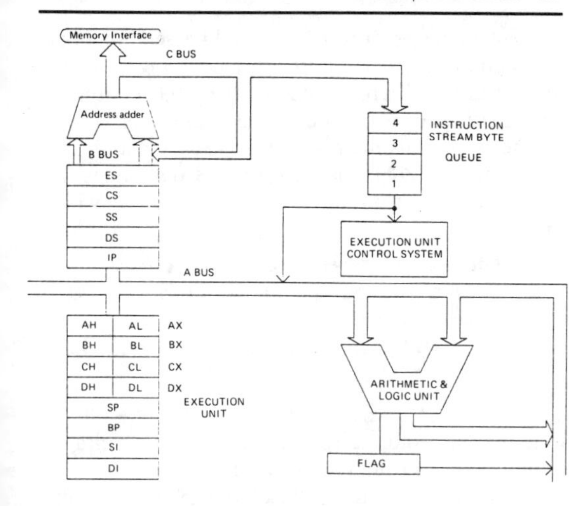

Figure 2.1 *A block diagram for the 8088 chips, showing the functional units and interconnections.*

number from one of four **segment** registers, shifting this number four places left (equivalent to multiplying the number by 16 or adding one hex zero to the end of the number) and then adding an offset 16-bit number to the resulting 20-bit number. The offset number will usually be contained in the IP register. This method of forming an address will be dealt with more fully in Section 3, and is known as **segmentation**, since each address can consist of a **segment address** and an **offset address**. On the 8088/8086 processors, each segment can be of up to 64K, and the use of the top 4 bits of the segment register allows up to 16 segments to be used, so that up to 1024K (1 Mb) of memory can be addressed. This scheme of segmentation is important, because even on the more complex processors which do not need to use segmentation (because they possess full 32-bit address registers) segmentation can be implemented so as to make the software of the 8088/8086 compatible.

The two sections can operate separately and asynchronously, but interact as required. The addressing section (BIU) deals with the fetching of instructions, queuing of instructions awaiting decoding, the fetching and storing of operands (data bytes and words), address relocation (jumps) and the control of buses. The data handling section takes instruction bytes from the queue and generates addresses which are handed to the BIU section for use in subsequent fetch or store operations.

The BIU deals mainly with the address and data buses. In the 8088, because only 8 data lines are required, the bus is made up from three parts which are:

- The higher address lines A8 to A15.
- The multiplexed data and address lines AD_0 to AD_7.
- The multiplexed address and status lines A16/S3 to A19/S6.

The A8 to A15 lines are output-only on the 8088, so that they are internally latched and their voltages remain valid throughout a bus cycle.

The instruction queue allows four bytes to be held pending transfer to the control unit for execution. Whenever there is room for a single byte in this FIFO (first-in, first-out) buffer, the BIU will attempt to fetch a byte from memory. If there are no bytes in the queue, any byte that is read from memory is shifted to the head of the queue to avoid waiting. The use of the FIFO buffer system ensures that efficient use is made of the buses, reducing the time ('dead time') when buses are idle during the processing of an instruction.

The action of the 8088 is necessarily slower than that of the 8086 because a word of 16 bits must be fetched in two 8-bit portions, entailing a time penalty of four clock cycles. The difference is more noticeable for a set of short instructions, because the use of the FIFO buffer will conceal the greater use of the buses when the more complex instructions are being executed. Because of the 8-bit organization, a word may be located starting at either an even or an odd address number without any penalty in fetch time.

The 8086

The block diagram for the architecture of the 8086 is illustrated in Figure 2.2. This is essentially similar to that of the 8088, as would be expected since the 8088 was developed from the 8086 rather than the other way round. The most important differences are:

- The use of multiplexed address/data lines AD_0 to AD_{15}.

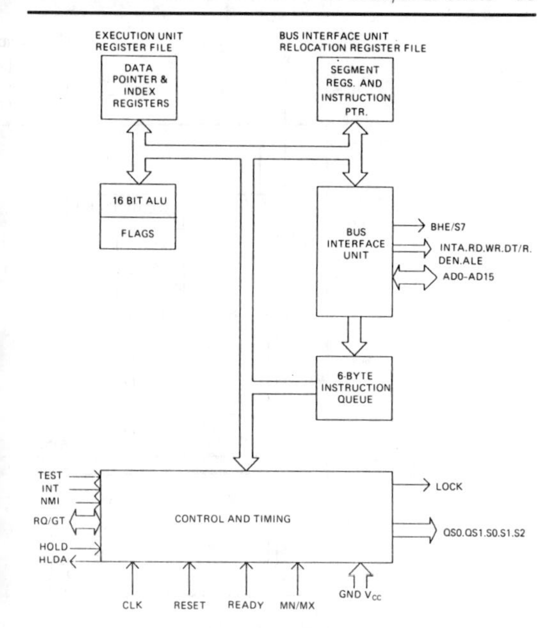

Figure 2.2 *A block diagram for the 8086. Though this is essentially the same as for the 8088, the diagram is differently arranged.*

- The presence of a bus high enable (BHE) signal to enable data to the most significant half (D_8 to D_{15}) of the data bus. This pin is not required on the 8088 because there is no upper data byte.
- Data is fetched and stored in words (two bytes).
- The use of a 6-byte (3-word) FIFO queue which will trigger a word fetch when **two** bytes are vacant.
- There will be a time penalty in fetching a data word from memory if the first byte of the word is located on an odd-numbered address. There is no delay in fetching instruction words, and the software does not have to take account of how the memory is used except in timing loops or other instructions where timing is important.

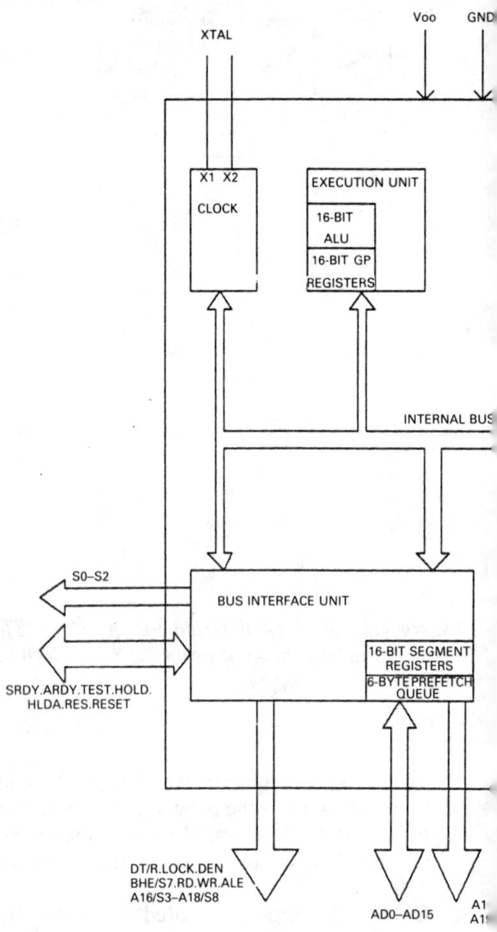

Figure 2.3 *The internal structure of the 80188/80186 in block form. The diagram is for the 80186, so that the separate address-only bus of the 80188 (for A8–A15) is not shown.*

B The 80186

The block diagram for the internal structure of the 80186 is illustrated in Figure 2.3. In this diagram, the units that are marked as execution unit and bus interface unit are functionally identical to the 8086,

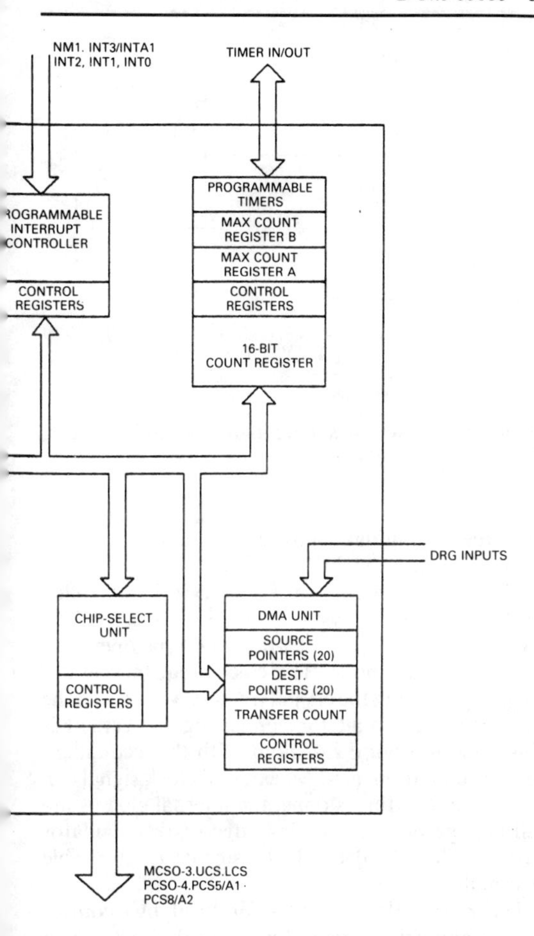

and the diagram concentrates on the relationship between this and the other units. These units are:

- clock generator;
- programmable interrupt controller;
- programmable timers;
- programmable direct memory access (DMA) unit;
- chip-select unit.

The actions of these sections will be dealt with only briefly here because Section 7 deals in more detail

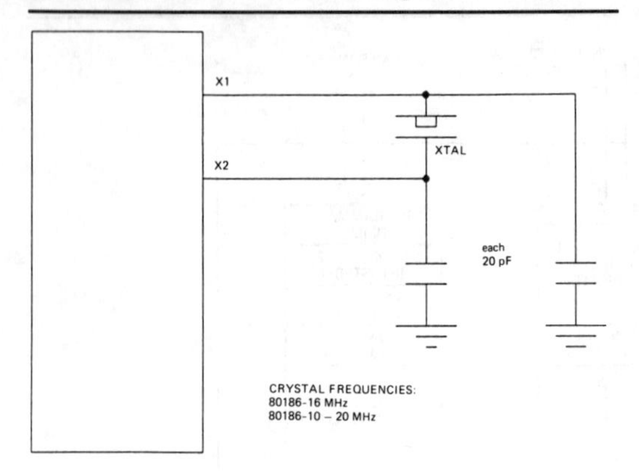

Figure 2.4 *The clock circuit for the 80188/80186, showing suggested components.*

with the individual chips whose actions are incorporated into the 80186.

The clock generator uses a crystal controlled oscillator, with the crystal externally connected. The crystal should be a parallel-resonant type operating in fundamental mode. The clock frequency is obtained by dividing the crystal frequency by 2, and the recommended circuit for connecting the crystal is illustrated in Figure 2.4, along with the recommended crystal parameters. An external clock signal can be used, and if this is done, the internal clock state will change on each trailing edge of the oscillator pulses. The divided clock signal is available externally.

The 80186 also provides for local bus control signals, along with signals that allow the local bus to be controlled externally. The usual read and write signals are generated, but there is no memory/IO form of signal. Type 8286/8287 transceiver chips can be connected for additional buffering and controlled by signals from the 80186, using the data enable and data transmit/receive pins. The use of HOLD/HLDA signals also allows one bus to be used at different clock frequencies by different units.

The various peripheral sections of the 80186 chip are controlled by the internal peripheral interface portion of the chip, using 16-bit registers in an internal 256-byte piece of memory. These 256 bytes can be mapped as memory addresses (starting at an address which is a multiple of 100H) or as I/O space. When the bus is using internal registers, the external address, data and control pins will be driven as they are during a normal action, but any data that is read will be ignored. The address block used by these registers can be altered.

The chip-select section of the 80186 allows memory and peripherals to be selected and for wait states to be generated. Address bits A1 and A2, which are latched, are also provided. Six chip-select outputs are available, with one each used for upper and for lower memory respectively, and four used to select in mid-range memory. The mid ranges can be set by software to 2K, 4K, 8K, 16K, 32K, 64K, 128K, and the lower and upper chip selects can be set for 1K and 256K respectively. The starting (**base**) address of the mid-range chip select can also be chosen, but only one chip select is permitted to be active for a given memory location at any instant. Chip selects can also be generated for up to seven peripherals.

The DMA controller of the 80186 is shown in block diagram form in Figure 2.5, and allows two independent DMA channels to be controlled, providing for the transfer of bytes to and from memory without passing through registers of the microprocessor. This can be used for memory-to-memory, or

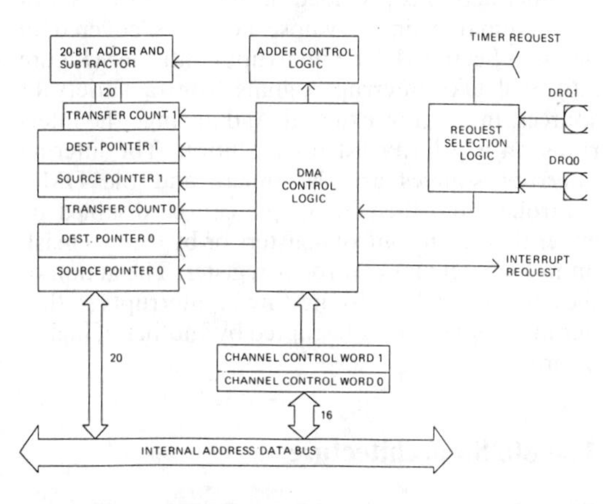

Figure 2.5 *A block diagram for the DMA controller of the 80188/80186 chips.*

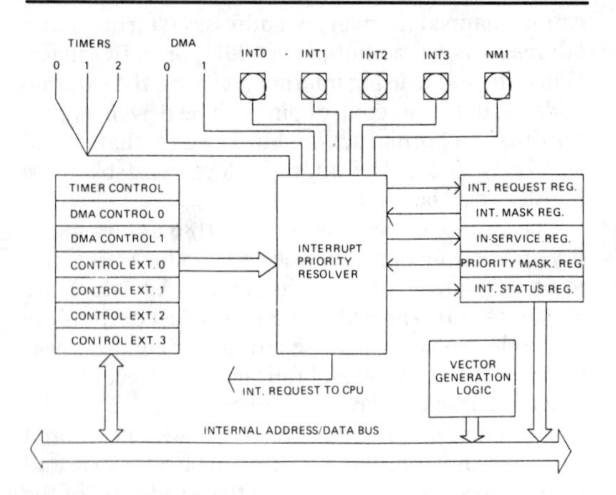

Figure 2.6 *The interrupt controller for the 80188/80186 in block form.*

memory to or from I/O transfers with data transferred as bytes or as words, using even or odd starting addresses when memory transfers are involved. The transfers make use of the 20-bit destination and source address pointer registers for each channel. The maximum transfer rate is 2.5 Mb/s at a clock rate of 10 MHz.

Timer action is provided in the 80186 by a set of programmable timers whose action is covered in detail in Section 4. The interrupt controller (Figure 2.6), will take interrupt signals from a variety of sources, internal or external, and arrange the interrupts on a priority list for attention. The internal interrupt sources are the timers and the DMA controller, and their interrupts can be disabled by either their own control registers or by using a mask bit in the interrupt controller register. The action of the controller allows for nesting of interrupts so that one interrupt can be interrupted by another of higher priority.

The 80286 architecture

The 80286 was developed from the 8086, but with a view to making much more of the capabilities of a 16-bit chip than was ever possible with the 8086. It was designed from the start to permit multiple-user and multitasking operation when used in its **protected** mode (so called because the chip design allows the work of any one user or task to be protected from the others), but in practice the chip has been utilized mainly in its **real** mode in which it emulates the 8086 and is compatible with applications programs written for the 8086, but operating at a much higher speed.

The block diagram for the 80286 is illustrated in Figure 2.7. The chip can be considered as consisting of four sections, the bus unit (BU), instruction unit (IU), execution unit (EU) and address unit (AU). There are two FIFO registers used as queues, one of 6 bytes (3 words) between the bus unit and the instruction unit, used to queue coded instructions, and one following the instruction decoder which contains up to three decoded instructions. The use of queuing allows for **pre-fetching** of instructions, so that at any instant one instruction is being fetched while the previous one is being decoded, and the instruction before that is being executed.

The address unit of the 80286 is considerably more complex than that of the 8086, and the whole address mechanism in protected mode is also different. Since the 80286 chip is never (well, hardly ever) used in protected mode by the MS–DOS operating system for computing applications, however, we will defer any detailed discussion of protected mode address formation until later, in connection with the 80386. As far as real mode is concerned, so that the 80286 can make use of software written for the 8086 and unmodified, the address unit allows the same scheme of segment register and offset to be used.

One important advantage is gained from the use of separate address and data buses as compared to the multiplexed operation in the 8086/8088. The 80286 allows for the **pipelining** of fetches from memory, meaning that a new address can be put on to the address lines while the processor is dealing with a previous instruction. This requires the address lines to be latched so that the address can be held on the bus. The use of pipelining along with instruction prefetch are the main factors in the much faster operation of the 80286 as compared to the 8086.

For protected mode addressing, the 80286 uses two registers. One provides an offset address number

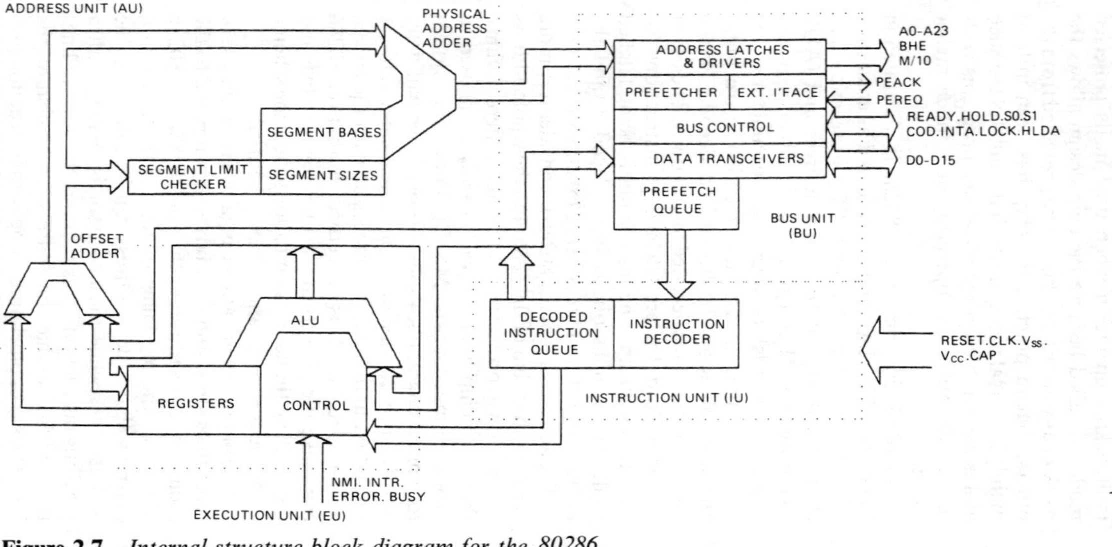

Figure 2.7 *Internal structure block diagram for the 80286.*

in the same way as is used on the 8086 chip. The other register, however, selects one of two **descriptor tables** in memory and selects a position in the descriptor table in which a segment number can be found. Each descriptor table can consist of up to 8192 entries, so that the two tables permit addresses for 16384 segments to be held. This amounts to a total of 1 Gb, and since the address pins do not allow for as much as this, much of this memory can be **virtual** memory which actually accesses the disk.

Since the operational codes for the 80286 are the same in protected mode as in real mode, the only barrier to running 8086 programs in the protected mode of the 80286 is that the address references in the programs need to be rewritten (and the descriptor tables set up ready to use), and if operating system calls are used, these have to be altered to suit an operating system that can cope with protected mode. This work has never been done, and programs, other than enhanced 'front-end' types, capable of using protected mode are more likely to emerge slowly for machines using the i486 (80486). Much of the demand for larger and larger memory space has arisen from the trend to making programs more 'user-friendly', and seems to be a self-accelerating effect.

The use of two descriptor tables in protected mode allows the 80286 to implement the protection system that is needed for multitasking. One descriptor table is a global descriptor table (GDT) whose segment addresses are available to any program (or **task**). The other table is a local descriptor table (LDT) which is used to contain descriptors that belong to a task – each task can have its own descriptor so that it occupies its own piece of real or virtual memory. Any type of descriptor other than interrupt and trap types (which use a separate table) can be put into the GDT, but the use of the LDT is more restricted.

The 80386DX and 80386SX

The 80386, 80386DX, 386 or i386 is a development of the 80286, building on the principles of addressing and pipelining in particular, but with a full 32-bit internal and external bus structure for the 386 chip, and with 32-bit internal but 16-bit external data bus

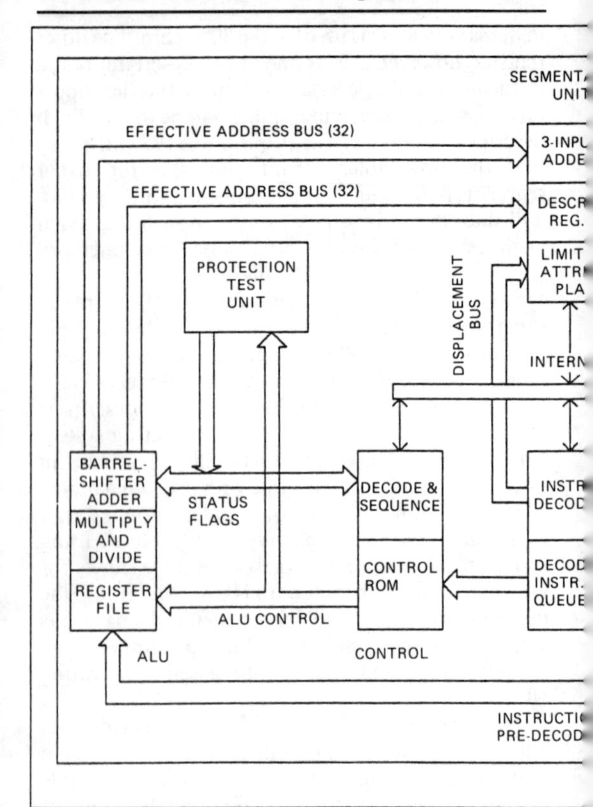

Figure 2.8 *The internal structure block diagram for the 80386DX.*

for the 80386SX. This allows the 80386DX the use of full 32-bit addresses and of 32-bit data types, though the applications of the 386 and 80386SX in computing have mainly been to much the same purposes as the 80286, running programs that were designed to run under MS–DOS on the 8088/8086 machines, but with the advantages of multiuser and multitasking abilities.

The block diagram of the internal structure is shown in Figure 2.8, and a notable feature is the number and size of internal buses. The chip can be imagined as composed of eight units which in

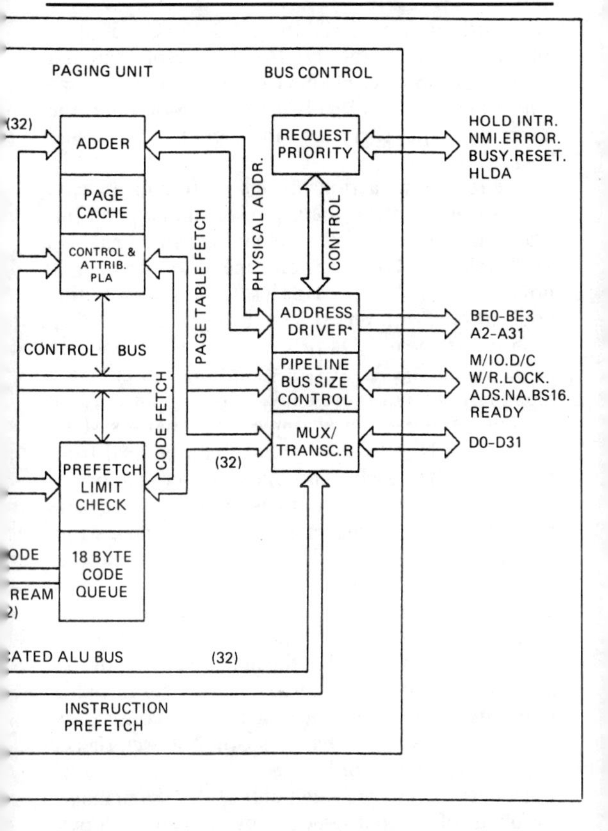

addition to carrying out the normal functions of any microprocessor incorporate address translation registers, memory management and memory protection.

The bus control unit consists of the request prioritizer, address driver, pipeline and bus size control and the multiplex and transceiver unit. The prioritizer deals with the interrupt, reset and bus availability signals, and is connected through an internal control bus to the portions that control the use of the main external buses, the address bus, data bus and other control buses.

The address bus uses lines A2 to A31 so that bytes can be read in blocks of four, but the use of four byte–enable pins allows individual bytes or other groupings of up to four bytes to be selected – this is dealt with in more detail in Section 6. The 32 data lines can be used either for 16-bit or 32-bit read and write operations, controlled by the voltage on one pin. When the data bus is used for 16-bit data, the upper data lines carry the same signals as the lower lines.

The remaining actions in this portion of the chip concern bus control, the use of pipelining, and bus size as mentioned above. Pipelining can be enabled or disabled, and the decision depends on whether or not the chip is being used (as it often is for computing applications) with very fast static RAM cache. When such cache memory is in use, there is no point in using pipelining. An attractive option is to use main memory in two or more banks which are **interleaved** so that, taking as an example a two-bank system, double-words (four-byte groups) can be read from the banks alternately. This allow slower memory to be used, and when pipelining is enabled there can be a considerable overlap of actions, with one instruction from one bank being decoded while another instruction from the other bank is being read. Figure 2.9 illustrates the methods that are used for two- and four-bank operation of the 386, using a 32-bit address bus and switching the banks by using the A2 and the A3 signals. The larger the number of banks that is used, the longer the time between consecutive accesses to the same bank, so that slower (hence cheaper) memory can be used.

The instruction pre-fetch unit of the 386 manages the queue of coded instruction bytes, using a 16-byte queue (four double-words). These 32-bit codes are then passed to the instruction decoder, and once decoded join another queue of three decoded instructions into the Control ROM section which is responsible for recognizing and implementing valid instructions.

The ALU is capable of 32-bit operations and is connected directly to the data-bus transceiver section by a 32-bit internal bus. In addition to arithmetic and logic operation on data, the ALU deals with address formation, sending address numbers over two 32-bit buses to the segmentation and paging units which will generate either physical addresses (capable of being used on a 32-bit address bus) or virtual addresses (numbers between 2^{32} and 2^{46}) as required.

The 80386 offers both segmentation and paging as

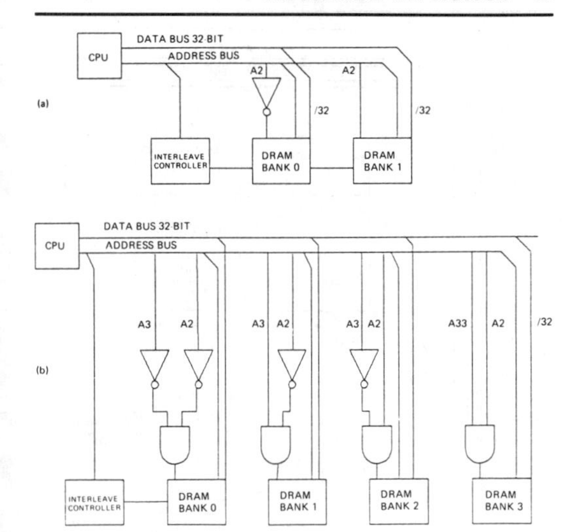

Figure 2.9 *Illustrating two (a) and four (b) bank memory interleaving.*

a way of dividing up memory. Unlike the 80286, the segments of the 80386 can be of any size up to the maximum possible physical memory of 4 Gb, but any practical system will use considerably smaller segments. The page is a 4K piece of memory, and a paged segment is a segment which is divided into 4K pages. Paging can be enabled or disabled to suit the way that the memory management is to be carried out.

The purpose of paging is to make the 'virtual 8086' operation of the 80386 chip easier. In virtual 8086 operation, the memory is divided into 1 Mb pieces, with a task running in one or more segments. Within one 1 Mb piece there will be a copy of MS–DOS and an application program, with its own stack, data and code, all of which is totally isolated from the rest of the machine. In two such tasks, it is likely that identical addresses will be specified in the programs, but these cannot correspond to identical physical addresses on the address bus of the 80386. Paging enables the conversion from an address specified in a program to an address on the address lines to be accomplished while keeping the different tasks separated.

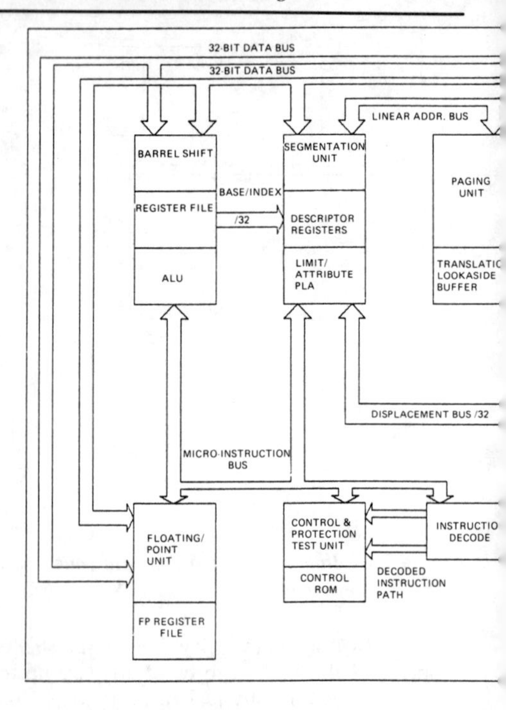

Figure 2.10 *A block diagram for the i486, showing the built-in cache and the floating-point unit.*

When paging is used, the address obtained from the segmentation unit is translated into a physical address for the address bus by way of a page table. The table consists of 32-bit entries, of which 20 bits will be used for the familiar 8086 type of address and 12 bits for page identification and status, including page protection. The collection of pages that a task uses is called the **repertoire**, and the repertoire table contains entries each of which consists of the address of a page table and the attribute bits that describe how the table will be used. The sequence of selection is therefore to translate an address number into a repertoire entry, which leads to an entry in the page table, which leads to a page in which the physical address is located.

This multiple indirection is potentially slow, and a cache system is used with RAM on the chip storing

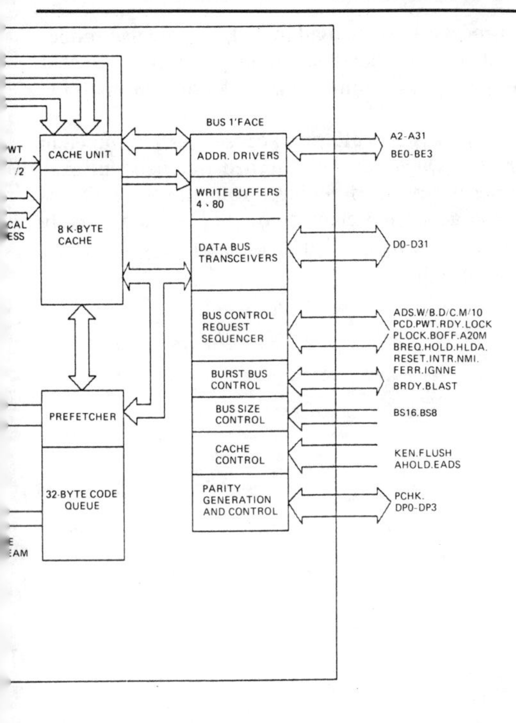

the 32 most recent page translations, so that the repertoire and page tables need be searched only if an address has not been translated previously. In the course of the use of paging by tasks, new translations will ideally be made only each 32 pages (a 128K memory space), and since a surprising number of 8086 applications programs use no long jumps (all code is confined in a single 64K space) or use only a few such jumps, this use of translation cache is very effective.

i486 (80486)

The architecture of the i486 chip is illustrated in Figure 2.10. The main points of difference from the 80386DX are the incorporated floating-point unit and the cache unit, both of which are completely integrated with the other units of the CPU. Segmentation and paging are implemented as on the 80386DX, using the same set of levels of protection.

The processor can be used in real or protected modes and also in virtual–8086 mode (within protected mode) to allow multitasking of conventional PC software.

The cache size is 8K, and keeps its data grouped in four sets which can be searched independently. A notable hardware difference from the 80386DX is that the generated clock frequency is used directly rather than being divided down internally as it is for the other chips.

3 Register models

8088, 8086, 80186, 80188

All of these processors use the same style of registers, whose functions can be grouped as data processing and address processing respectively. The address registers consist of one instruction pointer (IP) register and four segment registers identified as code segment (CS), data segment (DS), stack segment (SS) and extra segment (ES), as indicated in the table of Figure 3.1. All of these registers are 16 bits wide.

The process by which an address is formed is the nub of the addressing action of the 8086 set of processors, and the methods have spilled over into the 80286 and 80386 machines in order to maintain compatibility for applications programs running under MS–DOS. In addition, however, the organization of the addressing system in the 8086 family also makes it easy to recast the older programs that ran on the 8-bit 8080 chip, using CP/M, and which can be, and have been, converted to run on the 16-bit processors.

Address formation

One method of address formation makes use of the IP register only. Any program that was written for the 8080 chip running CP/M will use an address

-------IP-------	Instruction pointer, 16 bits
-------CS-------	Code segment, 16 bits
-------DS-------	Data segment, 16 bits
-------ES-------	Extra (data) segment, 16 bits
-------SS-------	Stack segment

The IP register is used in all addressing, with the upper four bits of each address taken from the segment registers. The CS register is used automatically for all instruction fetches.

Figure 3.1 *The address registers of the 8088/8086, used also on the 80188/80186.*

space of only 64K, and this range of addresses can be dealt with by using the IP register only, with the CS, DS, SS and ES registers all holding the same constant value throughout. For such a program, the only practical difference between the 8080 version and the 8086/8088 version is that the 8086/8088 version must run under MS–DOS, so that calls in the original program that were made to CP/M must be made to MS–DOS instead, and any direct references to screen addresses or addresses in the ROM BIOS must be rewritten. Programs of this type carry the extension COM and run in one segment.

Given that the IP register can address a space of 64K, the logical way of managing the memory is to divide it into 64K **segments**. The use of 20 address pins on these processors allows the use of 16 (the four upper bits of the address set) segments, and it would have been possible to make one 16-bit register serve for four different sets of segments by using its bits in groups of four. This however would have made the 8086 design a dead end, making further development impossible, so that the four separate 16-bit registers are used, along with a scheme for obtaining a 20-bit address by combining the content of a segment register with the content of the IP register.

The scheme consists of shifting the number in a segment register left by four bit places (equivalent to shifting one hex place left, or multiplication by 16 denary), and then adding this to the contents of the IP register. Suppose, for example, that the CS register contains the number which in hex is 4000H and the IP register contains 0100H. The left-shift action on the CS register produces 40000H, and when added to the IP number this gives 40100H, the effective address (EA) that will be used on the address lines. The shifting and adding is done in a 20-bit ALU, since the registers themselves are only 16 bits wide. The scheme is illustrated in Figure 3.2.

The segment registers will carry numbers such as 1000H, 2000H, up to F000H, for use as segment numbers, but the way in which address numbers are written can create considerable confusion. Figure 3.3 shows the printout from the DOSMAP utility, which shows the use of memory for resident programs, using full 5-digit hex numbers. Contrast this with the display in Figure 3.4, which uses the MS–DOS utility DEBUG to look at a part of the memory. The

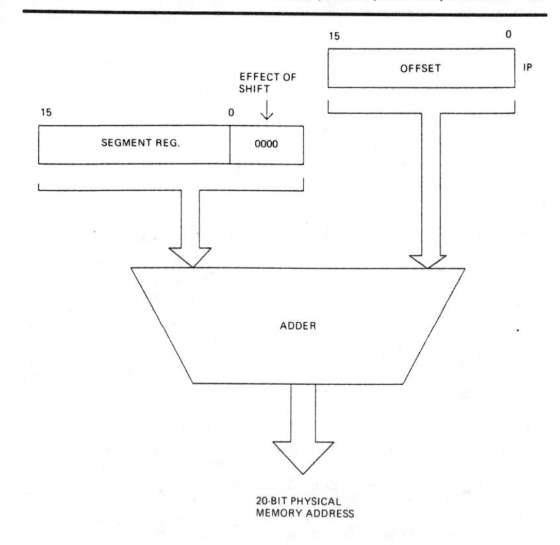

Figure 3.2 *How a 20-bit address is formed from the contents of two 16-bit registers. This method is available on all the chips from 8088 to i486.*

```
Start End    Length Segs Owner                                      Interrupts
-------------------------------------------------------------------------------
00000 097CF   38864   1  BIOS and DOS parameters and routines
097D0 0CE9F   14032   1  drivers + BUFFERS + FILES
0CEA0 0D88F    3312   2  resident COMMAND.COM                       22 23 24 2E
0D890 0D8CF      64   1  free
0D8D0 0DC3F     112   1  RTC.COM
0DC40 0E54F    2320   1  RTC.COM                                    09 EC F5
0E550 0E6EF     416   1  RTC.COM                                    21
0E6F0 0ECAF    1472   2  MARK.COM
0ECB0 1029F    5616   2  MOUSE.COM                                  06 08 10 33 F9
102A0 1930F   36976   2  SR.COM                                     13 2F
19310 9FFFF  552176   3  free
-\
```

Figure 3.3 *A print-out from the DOSMAP utility (from PDSL) showing address numbers in five-figure hex format.*

address that is shown for the first line in this display is 1020:02A0, which consists of the contents of the CS and the IP register in that order. When we use the rule of shift and add, this becomes 10200+02A0=1022A0, and this could also have been written as 1000:22A0, identifying this address as segment 1 and IP (**displacement** or **offset** address) 22A0. Because of the way that an address is formed,

```
1020:0240  C0 00 BF 00 C9 1F 70 00-CC 18 70 00 E8 8F 00 00   ......p...p.....
1020:0250  E4 05 5C 0F 40 01 00 02-0A 0D 80 01 00 3A 49 19   ..\.@.........:I.
1020:0260  11 03 11 03 00 00 00 00-03 FF 00 FF 00 00 2B 80   ..............+.
1020:0270  FA 80 74 0A 80 FA 81 74-05 2E FF 2E 28 01 FB 53   ..t....t....(..S
1020:0280  51 52 56 57 1E 06 8C CE-8E DE FC 80 3E 49 01 00   QRVW........>I..
1020:0290  74 39 80 FC 02 74 3D 80-FC 03 75 03 E9 98 01 08   t9...t=...u.....
-d
1020:02A0  E4 74 1E 80 FC 05 72 23-80 FC 07 76 14 80 FC 09   .t....r#...v....
1020:02B0  74 0F 80 FC 0B 74 0A 80-FC 0C 74 7E 80 FC 0D 75   t....t....t~...u
1020:02C0  0A 50 C6 06 4B 01 00 E8-EB 01 58 07 1F 5F 5E 5A   .P..K.....X.._^Z
1020:02D0  59 5B EB A5 E8 92 02 A0-36 01 30 E4 01 06 04 01   Y[......6.0.....
1020:02E0  83 16 06 01 00 E8 64 02-75 23 80 3E 4A 01 00 C4   ......d.u#.>J...
1020:02F0  3E 3C 01 1E 56 C5 74 08-75 16 B9 00 01 F3 A5 5E   ><..V.t.u......^
1020:0300  1F 89 3E 3C 01 E8 69 01-75 DB E9 2D 00 E9 36 00   ..><..i.u..-..6.
1020:0310  B9 40 00 BB CD AB AD 01-C3 AB AD 01 C3 AB AD 01   .@..............
-d
1020:0320  C3 AB AD 01 C3 AB E2 EE-39 1C 74 D3 5E 1F C6 04   ........9.t.^...
1020:0330  FF FF 06 18 01 E8 A0 02-EB 0C 31 C0 07 1F 5F 5E   ..........1..._^
1020:0340  5A 59 5B CA 02 00 A0 36-01 30 E4 01 06 0C 01 83   ZY[....6.0......
1020:0350  16 0E 01 00 E8 CB 00 72-E3 E8 70 03 56 E8 EF 01   .......r..p.V...
1020:0360  75 0D E8 53 02 5E 81 06-3C 01 00 02 E9 04 00 5E   u..S.^..<......^
1020:0370  E8 07 00 E8 FE 00 75 E1-EB C0 C4 3E 3C 01 26 8A   ......u....><.&.
1020:0380  05 26 F6 15 3A 04 26 3A-05 AA 75 06 C6 06 4D 01   .&..:.&:..u...M.
1020:0390  FF C3 56 B9 0A 00 30 C0-EB 03 83 C6 0C 38 04 E0   ..V...0......8..
```

Figure 3.4 *A print-out from the DEBUG utility (part of MS-DOS) which shows address numbers in CS: IP form.*

the same address can be written in many different ways, and some utilities, like DEBUG, will present an address in a way that requires interpretation such as is illustrated above. This is because DEBUG follows a convention of locating the first program byte of a file and assigning it to the IP number 0100, then calculating the CS number to suit. This is an inheritance from the COM type of program as used under CP/M running on the 8080 chip. Such a program will keep its code, its data and its stack within the same 64K segment.

The type of program that is given the EXE extension will in general use more than one segment. The DS register can carry the number of a segment that will be reserved for the data of the program, and the SS register can carry the number of a segment reserved for the stack. The extra segment, if used at all, is used mainly as an additional data segment, particularly if data is being moved from a space in one segment to a space in another. In general, the operating system and memory-resident programs (see Section 12) will take up the first two segments 0 and 1, and since only 10 (denary) segments of RAM are used in machines running MS–DOS, this leaves 6 segments available for code, some 384K. This sets the effective upper limit of size of a program running under MS–DOS. The top 6 segments of memory are reserved for use by ROM and by RAM dedicated to the video display, though bank-switching can be used to make use of RAM at some address ranges, a scheme known as **expanded memory**.

Within any program that makes use of more than one code segment, program instructions that specify an address must be capable of specifying a full address using the CS segment and the IP register. The instruction set therefore allows for both short and long addresses to be used. A short address changes only the content of the IP register, working within one segment; a long address reference changes both CS and IP register allowing for changes of segment.

The data registers

The data register map, including flags, of the 8086/8088 family of chips is illustrated in the diagram of Figure 3.5. There are four general registers designated as AX, BX, CX and DX, plus four

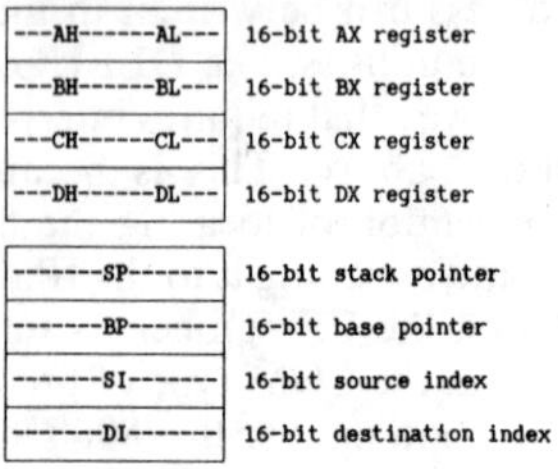

```
|---AH------AL---|  16-bit AX register
|---BH------BL---|  16-bit BX register
|---CH------CL---|  16-bit CX register
|---DH------DL---|  16-bit DX register

|-------SP-------|  16-bit stack pointer
|-------BP-------|  16-bit base pointer
|-------SI-------|  16-bit source index
|-------DI-------|  16-bit destination index
```

Figure 3.5 *The data register map for the 8088/8086, applying also to 80188/80186.*

pointer and index registers SP, BP, SI and DI. As before, all of these registers are 16 bits wide.

The pointer and index registers are used in forming 16-bit indirect addresses. Any address reference used by an instruction can use a number that is built up by adding any combination of a base register number, an index register number and a displacement number. The BX and BP registers can be used to contain base numbers, and the SI or DI registers can be used to contain index numbers (see Section 8). The SP register is intended to be used by the stack pointer, and it has the same relationship with the stack segment register SS as the IP register has with the CS register.

The general-purpose registers are intended for the manipulation of data, and each of them can be used either as a single 16-bit register or as a pair of 8-bit registers, so allowing for compatibility of 8080 codes. When the 8-bit units such as AH and AL are used, complete separation is enforced – an addition in AL that generates a carry, for example, does not place the carry bit automatically into AH. All four of these general-purpose registers can be used in this way, providing eight 8-bit registers if needed; a useful feature for programs that work with the ASCII codes of text.

Ideally, all four of the general-purpose registers would be completely interchangeable. This, however, does not lead to the most efficient use of registers because it would add an overhead in the form of a register reference code to many instructions which take long enough already. The AX register, following the pattern of the 8080 processor, is the most widely used register for logic and arithmetic work. The BX register is available for a variety of general actions, but is also specified as a base register so that in base-addressed memory references the number in the BX register will be added to the displacement number specified for the instruction.

The CX register is used by many counting applications, so that several instructions of the 8086 assume that a count number will be present in CX. Loops, for example, can check for a number in CX as well as for any explicitly-stated conditions. The CL half of the register can be used in rotate instructions to specify how many bits are to be rotated. Many of the arithmetic instructions, particularly for multiplication, division and input/output, are also specific to registers AX or to AX and DX.

The flag register

The flag register, usually referred to as the F-register but sometimes called the status register, isn't really a register like the others. The bits in this register can be tested, but little else and they don't fit together as a number. Each of the 9 bits used in the register records the effect of the previous step of the program. If the previous step was a subtraction that left a register storing zero, then one of the bits in the flag register will go from value 0 to value 1 to bring this to the attention of the CPU. If you add a number taken from memory to the number in an accumulator, and the result consists of seventeen bits instead of sixteen then another of the bits in the flag register is 'set', meaning that it goes from 0 to 1. If the most significant bit in a register goes from 0 to 1 (which might mean a negative number), then another of the flag bits is set.

Each bit in the flag register, then, is used to keep a track of what has just happened. What makes the flag register so important is that you can make branch commands depend on whether a flag bit is set (to 1) or reset (to 0). Figure 3.6 shows how the bits of the flag register of the 8088/8086 are arranged. Of these bits, numbers 0, 6 and 7 are the ones that are most used. Bit 0 is the carry flag. This is set (to 1) if a piece of addition has resulted in a carry from the most significant bit of a register. If there is no carry, the bit remains reset. When a subtraction is being

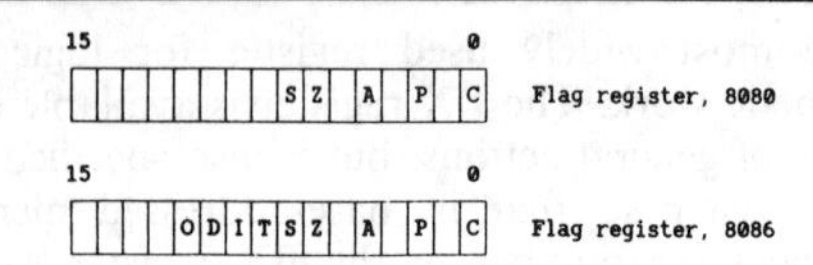

Figure 3.6 *The flag register bits of the 8088/8086. Note that the 8086 uses four flags that are not available on the 8088. The same bits apply to the 80188/80186.*

carried out (or a similar operation like comparison), then this bit will be used to indicate if a 'borrow' has been needed. It can for some purposes be used as a seventeenth bit for the accumulator, particular for rotate operations in which the bits in a byte are all shifted by one place. The carry bit is used by all of the addition and subtraction operations of the 8088/8086, no matter what registers are being used.

The zero flag is bit 6 of the flag register. It is set if the result of the previous operation was exactly zero, but will be reset (0) otherwise. It is a useful way of detecting equality of two bytes – subtract one from the other, and if the zero flag is set, then the two were equal. The CMP (CoMPare) action will set or reset this flag without actually carrying out the subtraction action. The sign flag, number 7, is set if the number resulting in an action in a register has its most significant bit equal to 1. This is the type of number that might be a negative number if we are working with signed numbers. This bit is therefore used extensively when we are working with signed numbers.

The remaining flags are O, D, I, T, A and P, of which the O, D, I and T flags are used on the 8086 only. The O(verflow) flag is set when an arithmetic overflow has occurred in signed arithmetic, indicating a possible error. The D(irection) flag is used in data transfer instructions to indicate whether addresses are incremented or decremented after each read. The I(nterrupt) flag is used to enable or disable the maskable interrupts (see Section 4) of the processor, and the T(rap) flag can be set to force the processor to generate an interrupt after executing each instruction so that a program can be checked by **single-stepping**, executing each instruction and then pausing to examine register contents. The A(uxiliary carry) flag is used in BCD arithmetic to report a carry out of a 4-bit part of a register, and the P(arity) flag will be set or reset according to the parity count of bits in a byte, odd or even.

The flag register is affected by a few specialized instructions. The LAHF instruction, code 9FH, copies the flag register into AH, and the SAHF instruction, code 9EH copies the AH register to the flag register. This lattter instruction allows a program to alter the flags. The other two flag-specific instructions are PUSHF (9CH) and POPF (9DH) to push the flag register on to the stack and pop it back respectively.

Note that several bits of the flag register are not used, and are indicated as Intel reserved. These bits are likely to change state after some instructions, but not in a logical pattern. You should never write any software that depends on assuming the state of any of these bits; they must always be masked out.

One peculiarity of the 8088/8086, which was carried over from the old 8080 is that only certain actions, mainly actions that affect a register, will cause flags in the flag register to be affected. This takes a lot of getting used to if you have ever programmed some other types of microprocessors, particularly the 6502 or 6809. In particular, load and store operations never affect flags in any way, so that a flag which has been set before a load or store operation will still be set after that operation. This can at times be very useful.

The 80286 registers

The register map for the 80286 follows the same pattern as that of the 8086 with four additions to the registers whose contents are accessible, one of which is a machine status word (MSW) register. The use of this register and the flag register is illustrated in Figure 3.7, showing that only the lowest four bits are used, the remainder being Intel reserved. In addition, three more bits of the flag register are used for purposes specified to the 80286. The flag register bits 0 to 11 follow the same pattern as those of the 8086. The other additional register is the task register.

The additional flags are intended to deal with the

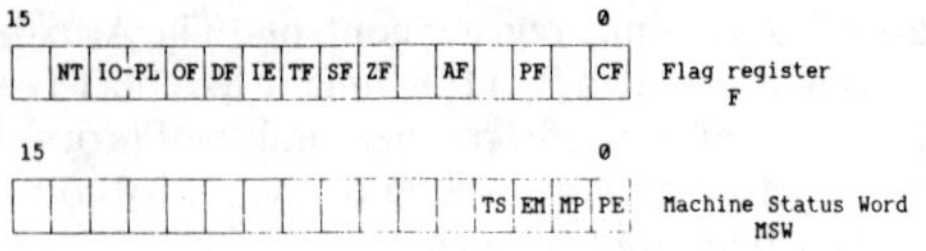

15 0

	NT	IO-PL	OF	DF	IE	TF	SF	ZF		AF		PF		CF

Flag register
F

15 0

											TS	EM	MP	PE

Machine Status Word
MSW

Bit spaces shown blank are reserved - when testing either of these registers always mask reserved bits, because some may be set or reset at random.

Figure 3.7 *The flag register and machine status word (MSW) for the 80286.*

use of the 80286 in protected mode. Recall that the 80286 can either emulate the 8086 (real mode) or can make full use of its larger address bus for multiuser and multitasking applications in protected mode. The instruction set of the 80286 provides for copying the MSW to a register or copying the content of a register to the MSW.

This latter action contains a significant prohibition. When the MSW bits are changed by copying a register to the MSW, any attempt to change bit 0 of the MSW from 1 to 0 will be ignored. This makes it impossible to return from protected mode to real mode in any simple way, such as might be required if a program running in protected mode required the use of an operating system action that was obtainable only in real mode (such as disk access). The 80286 cannot emulate the action of the 8086 in protected mode, and only a return to real mode can allow such emulation. This can be achieved only with hardware that will start a system reset and change bit 0 on the MSW during the reset cycle. This has made the 80286 a chip used only as a fast 8086, and practically never for the protected mode work that was envisaged for it at the design stage, though the OS/2 operating system accomplishes the switching in order to make full use of the 80286.

Dealing first with the additional flag register bits, bits 12 and 13 are used to record the I/O privilege level in protected mode. This allows four levels, 0 to 3, of privilege to be assigned, with level 0 being the highest privilege level. Each segment can have its own privilege level (held in a 'descriptor table') which can be transferred to the MSW, and which will determine access to other parts of the memory. Level 0 is an overall (supervisor) level which allows access to any segment, whereas any program working in a segment with level 3 privilege can gain access only to other level 3 segments, never to the higher levels.

Bit 14 in the flag register is the nested task (NT) flag. This is set when a task is being carried out nested inside another task. If this bit is set, there will be an entry in the task state segment (see later) that links back to the previous task. This applies only to protected mode operation.

The four bits in the MSW register start with the protection enable flag in bit 0. This bit is zero when the processor is started up, and this has the effect of putting the 80286 into real mode (8086 emulation), allowing disk reading under MS–DOS. Placing a 1 into this bit position will put the chip into protected mode, from which it cannot be changed except by a RESET action.

Bit 1 in the MSW register is the monitor process bit. When this bit is set (to 1), it will detect a WAIT instruction, and generate an interrupt number 7, the **processor extension not available** exception (see Section 4). Processor extension in this context means a co-processor such as the 80287, and this bit will be set when one task has been switched while the task was using the co-processor. When bit 2 is set, an ESC instruction will cause a **processor extension not present** exception, but allow emulation of the co-processor action.

Bit 3 of the MSW is the task–switch bit which is set each time the processor switches from one task to another. If this and bit 1 are both set, then the **processor extension not present** exception will occur for either the WAIT or the ESC commands.

In each case where the interrupt No. 7 has been invoked by these flags, this will allow software to test for the presence of a co-processor and to find if the co-processor is still working on computations from a different task. There is no problem about gaining access to a co-processor (or emulating 80287 action in software) after a task switch if no previous work remains being processed.

80286 addressing and descriptor registers

The 80286 in its real mode forms addresses in exactly the same way as is used for the 8086, combining the contents of two 16-bit registers so as to form a 20-bit address which is put out over the lower 20 lines of the address bus. This mode ensures total compatibility with the 8086 so that software which runs under MS–DOS on the 8086 will run on a machine using the 80286.

The 24 address lines of the 80286 can be used in real mode only if software is written to make use of the memory beyond 1 Mb. Such memory is referred to as **extended memory** (not to be confused with the bank-switched **expanded memory**) and any software that makes use of extended memory is incompatible – it cannot be run on the 8086 machines. Because of this, it is more common to use software that allows extended memory to be used as expanded memory by software that must run on either type of machine.

The full addressing capabilities of the 80286 are obtained only when the chip is used in **protected mode**. At this time, when the 80286 is used only as a faster version of the 8086, the main reason for describing the protected mode of the 80286 is as a way of introducing the addressing modes of the 80386. The system has to satisfy the requirements of using the full 24 address lines of the chip, and also of allowing separate programs (called **tasks**) to run in different parts of the memory. This requires a much more elaborate system in order to prevent tasks affecting each other and to ensure that it is possible to move from one task to another and still resume an earlier task without loss of any data.

The basis of protected mode addressing is that addresses are formed indirectly. In the 8086, each address is formed from numbers in the registers of the CPU. The 80286, by contrast, uses the numbers contained in the registers to refer to look-up tables in the RAM, from which the addresses are obtained. This allows the machine to create a range of addresses for each task it is running, and so ensure that there is no conflict of addresses even when two different tasks require the same internal address numbers. For example, two tasks might contain a call to 3FF5BH, but if the programs are loaded into different 1 Mb spaces in the memory, these memory calls to **program-relative** or **logical** addresses have to be translated into the correct **physical** addresses in order to locate the correct portions of memory. In addition, each task must preserve a record of its CS, DS and SS segments in RAM so that this can be used to ensure correct entries into the corresponding CPU registers when a task is run.

The entry into RAM that holds these memory details is called a **descriptor table**, and two such tables, a **global descriptor table** (GDT) and a **local descriptor table** are used. The 80286 has two additional registers, the GDT and the LDT registers, to deal with these address tables. The access to these registers is very limited, and obtainable only at privilege level 0. Entries into these registers are normally made by the system as tasks are loaded. The main reason that MS–DOS programs cannot use the 80286 in protected mode is that there is no provision in such programs for loading the descriptors, since such provisions are not compatible with the 8086.

The registers for each table consist of a 24-bit register for the base address at which the descriptor table starts, and a 16-bit limit register to indicate the position of the end of the descriptor table. An interrupt No. 13 will be generated if any attempt is made to find a descriptor at a position outside the limits. There are instructions for loading the global descriptor table register and for loading the local descriptor table register with both base and limit numbers. There is also a register for the interrupt descriptor table which can be similarly loaded and which contains the address and limits for the interrupt descriptor table. The loading instructions will each load a 6-byte number, consisting of 5 bytes for the table and limit numbers, plus one reserved byte.

The use of the GDT and the LDT allows the existence of separate protected tasks. The GDT contains descriptors which are available to all tasks, and it can contain descriptors of all types except interrupt and trap types which must be private to each task. The LDT can use only segment, task gate and call gate descriptors.

A segment descriptor consists of 4 words (8 bytes) arranged as in Figure 3.8 with the whole of the most significant word reserved. This most significant word will always be 0000H for the sake of compatibility with the 80386. The segment descriptors can be of four types, code, data and stack, reflecting the CS, DS and SS registers of the CPU, and system segment which hold the execution state of a task. Taking a typical code or data segment descriptor, the lowest order word is a limit word which gives the physical size of the segment – this must obviously be no more

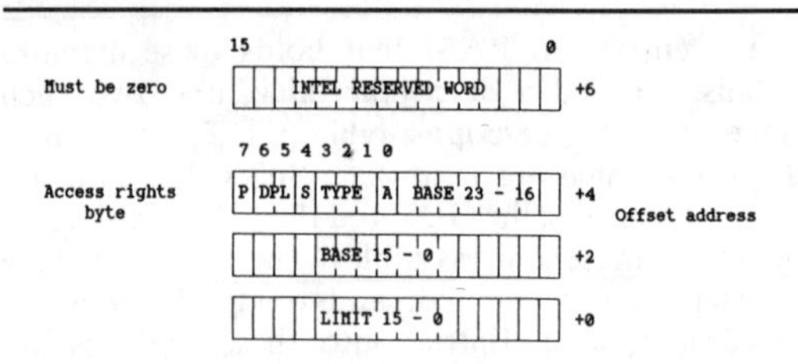

Figure 3.8 *The layout of a segment descriptor for protected mode use.*

The access rights byte consists of:
P - present DPL - descriptor privilege level
Type bits E (executable) ED (expansion direction)
W (writeable) C (conforming) R (readable or A (accessed)

Bit	State	Action
7	P=1 P=0	Segment is mapped into physical memory space Virtual, base and limit not used
6-5	DPL	Privilege attribute bits used for testing privilege
4	S=1 S=0	Code or data or stack segment descriptor System segment or gate descriptor
3 2 1	E=0 ED=0 ED=1 W=0 W=1	Bits 2 and 1 refer to data segment descriptor - Expand up segment, offsets must be less than limit or equal Expand down segment, offsets must be more than limit Data segment must not be written Data segment may be written
3 2 1	E=1 C=1 R=0 R=1	Bits 2 and 1 refer to code segment descriptor - Code segment executable only if code privilege level ≥ data privilege level, and code privilege level unchanged. Code segment may not be read Code segment may be read
0	A=0 A=1	Segment has not been accessed Segment selector has been loaded into segment register or used by selector test instructions

Figure 3.9 *The access rights bits and their meanings.*

than 64K, and it will be normal for this word to indicate a full 64K segment size.

The next word contains the lower 16 bits of the full physical address of the segment in the main memory, and the lower byte of the third word contains the upper 8 bits. This amounts to 24 bits, the full capacity of the address lines of the 80286, defining where this segment is stored. The upper byte of this third word contains a set of **access rights** bits whose functions are summarized in the table of Figure 3.9.

Four of the access rights bits indicate the type of segment. Code segments are always read-only, but data segments can be set as read-only or as read/write and will grow upwards, meaning that as

```
Access byte for system segment descriptor, byte 5 (word 2)

Bit(s) Name  Value              Action
0-3    TYPE    1    Available task state segment (TSS)
               2    Local descriptor table
               3    Busy task state segment
 7      P      0    Descriptor contents not valid
               1    Descriptor contents are valid
5-6    DPL    0-3   Descriptor privilege level

Base and limit numbers are coded in bytes 0 to 4 as for code/data descriptor.
```

Figure 3.10 *The access rights bits for a special segment descriptor.*

data is added higher memory addresses will be used. Stack segments must be read/write, and will normally grow downwards. The conforming bit applies to code segments only and when this bit is set execution of code is possible only when the privilege level for the code is greater than or equal to the privilege level for the data. This allows the code to be used for data of programs that have different (but not higher) privilege levels. The lowest-order bit records whether there has been any access to the segment, and the highest-order bit determines whether the segment is located in real physical memory or in virtual (disk) memory.

The system segment descriptors are for special-purpose descriptors. One use is to keep a descriptor for a local descriptor table itself, the other use is to keep track of segments which contain a partly-completed task. The special segment descriptors follow a pattern that is virtually identical to that of the other segment descriptors, but the higher-order byte of the third word is an access byte of a slightly different type. The use of the bits in this byte is illustrated in Figure 3.10, with four bits used to define type, two for privilege level, and one for validity (which includes being present in physical memory as distinct from being in virtual memory). Bit 4 is always zero to signal that this is a system control descriptor.

The type bits will indicate that the descriptor is for an available task state segment, a local descriptor table, or a busy task state segment. Privilege level applies only to the task state segments, which are segments that store the entire current status of a task. This will include contents of all processor registers, the address space, and links to the previous task. A jump or call instruction to a task state segment or to

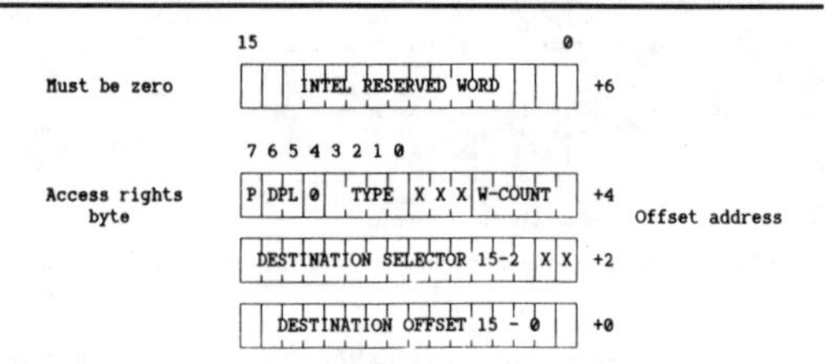

Figure 3.11 *The general form of a gate descriptor.*

a task gate descriptor is the method that is used to change from one task to another.

Gate descriptors contain addresses of entry points that a task can use, and there are four classifications, call gates, task gates, interrupt gates and trap gates. Call gates are used to reorganize the calls to the operating systems that would be used in a simple 8086 MS–DOS system, with the important difference that the use of a call gate can carry out protection checks and also determine the address at which the operating system routine is entered. Protection checks are used to ensure that programs operating at a low privilege level do not gain access to operating system routines that can alter programs at higher privilege levels, or change the privilege level of the existing program.

A call gate can be used by a program if the privilege level of the code of the call gate is equal to or lower than the privilege level of the calling program. The use of a call gate allows a routine at any level of privilege to be called – but only if a call gate has been set up for it. A call gate can even be used to call routines which exist at the same privilege level and even in the same memory space as the calling program.

The general form of a gate descriptor is shown in Figure 3.11. The lowest order word is the destination offset, which will be the address relative to the segment that is used for the code to which access is required. The next word is the destination selector which finds the correct segment – the lowest two bits of this word are not used. The third word contains some bits that do not correspond to anything on other gates, along with the familiar two bits for descriptor privilege level, four bits for type and the presence bit which determines whether the address words are valid.

The lowest five bits of the word are used as a 'word count', and only when the gate type is a call gate. They allow a count of up to 32 parameter words to be copied from the stack of the calling program to the stack of the called routine when the codes are at different privilege levels. A call gate may be situated in the global descriptor table (so that it can be accessible to any routine, though only routines with the correct privilege levels can use the gate) or in the local descriptor table, which makes the gate usable only by the local routine.

Interrupt gates and trap gates will be located in the interrupt descriptor table. An interrupt gate will be used in response to an interrupt when a program is running and will provide the address for an interrupt service routine. A trap gate is used for an exception (see Section 4), and will contain the address of the routine that services the exception. The same form of gate descriptor is used, but the bits of the word count are irrelevant.

Task gates are slightly different. Call gates, interrupt gates and trap gates all provide addresses into code segments, because code routines are being called by each type of gate. A task gate is called when a task switch is to be made, so that the call will be to a task state segment, and the destination offset word is not used because the task to which the machine is being switched will resume at whatever state was stored earlier in the task state segment.

Summary of 80286 protected-mode addressing

The addressing system of the 80286, compared to that of the 8086, is considerably more convoluted, so that a summary of what goes on during the loading and running of a program will help to clarify it. Assuming that you have an 80286 running in a computer with a suitable operating system, the program will be loaded from disk using the routines of the operating system, which will then read the program parameters and set up the descriptor tables, particularly the local descriptor table for the program. Several other programs may be loaded in, each with its own local descriptor table, and the operating system will use task switching as appropriate if more than one program is to be run at the same time.

When a program task is running, the code segment register in the CPU will contain the number that gives access to the segment descriptor. This will supply a 24-bit segment base address to the CPU, and to this address will be added the offset provided by the IP register in the CPU. The resulting full 24-bit address will be put out over the address lines. When data has to be used, the CPU will activate the data segment register, which will select a descriptor and provide another 24-bit base address in the same way, and when stack actions are required, the stack segment register will operate in the same way, finding a descriptor and causing an address to be loaded.

All of this suggests that the normal amount of alternation among code, data and stack could be cumbersome, involving a load from memory. It would be, and therefore the 80286 uses a set of registers as cache memory for segment descriptors – these are, inevitably, described as cache registers. The first time a segment register is used, its action will find the descriptor in the table and this descriptor will be loaded also into the cache register for the calling segment, so that an entry into the code segment register of the CPU will load three bytes of a descriptor into the CS cache register.

From then on, for as long as the same word is present in the CS register in this example, the cache register for CS will remain unchanged, and the base address that is to be used will be read from this copy of the descriptor. After a few instructions have been processed, the cache descriptors for CS, DS, SS and ES will be occupied, and will change only when a new segment number is selected – something that is likely mainly for the code segment. The use of these cache registers is entirely automatic, and there is no provision for any access to these registers by the programmer.

In the course of running the program, calls will have to be made to the operating systems for disk access, use of the VDU and other functions. These calls will be made by way of the gate descriptors, using the gate descriptor tables that will have been set up as the program was loaded. Calls to disk, VDU and keyboard are so universal that these are likely to be global and located in the global descriptor table. Some other calls will not be provided, or will be provided only locally, or will be at the highest level of privilege, inaccessible to any user other than a supervisor who has a suitable program. Interrupts will be dealt with by the interrupt gate addresses, and exceptions (program errors) by the routines obtained by way of the trap gates.

Unless the machine is running only one program, however, there is likely to be at some point a task switch. All of the parameters of the task that is currently running will be saved in a task segment, and the task switch gate will be used to move to the new set of segments where the new task is running. The parameters for this task will be taken from its task state segment, and it will resume from wherever it reached before a previous switch – in some cases this might only have been a holding loop for a program that was awaiting an input from the user. This program will use some of the global descriptors, but it will mainly make use of its own local descriptor table, which will maintain the separation between it and the other programs that are running. This program will then make use of its CPU segment registers as described earlier, but translating these numbers into quite different addresses in the physical memory. When another task switch occurs, the parameters will be stored in the task state segment and a new task, or another old one, will run.

All of these actions will normally make use of the 16 Mb of physical memory that can be addressed by the 24 address lines. If a call is made to a descriptor which contains an address that is not physically realizable, this is treated as a virtual address. Remember that the segment descriptor allows for the selection of 65536 segments which can each be of up to 64K in size. This corresponds to an address limit of 4 Gb (1 Gb = 1000 Mb), which is why the limit word is used to indicate how much memory is physically available. When an amount beyond the limit is selected, an exception is generated and it is up to the operating system to provide a routine to deal with this exception. This usually consists of a translation of the addresses into a disk sector coding so that memory space on the hard disk can be used.

80386 registers

The main reason for devoting so much space to the registers and protected addressing system of the 80286 is that it serves to introduce the ideas that have

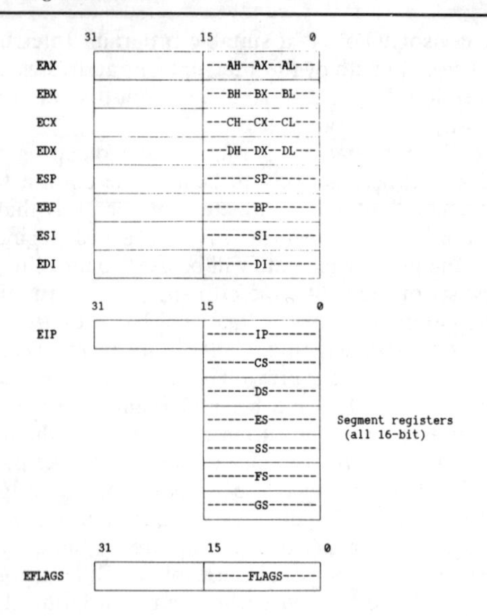

Figure 3.12 *The main register map for the 80386DX.*

been pursued rather further on the 80386 and 80486 chips. It is most unlikely now that 80286 chips will be used in the protected mode, because it makes much more sense to make use of the better design of the 80386 chip. This does not make the 80286 in any way undesirable, because most computer users simply want to run 8086 applications faster, and this applies to 80386 machines also. No user is volunteering to be the guinea-pig for a new and untried operating system for which little or no software exists. Note that the same prohibitions on use of Intel reserved register bits holds with considerably more force for the 386. These bits must be masked out when a register is tested, and must not be used for any storage purposes.

The main register map for the 386 is illustrated in Figure 3.12. The general data and address registers consist of 32-bit versions of the registers of the 80286, and the lower word of each can be used separately in the same way as it would be on the older processors. The letter E is used to indicate a 32-bit extended register, so that AX refers to a 16-bit register, EAX to a 32-bit register of which AX is the lower word. The

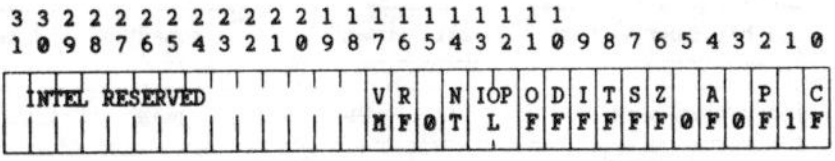

```
3 3 2 2 2 2 2 2 2 2 2 2 1 1 1 1 1 1 1 1 1 1
1 0 9 8 7 6 5 4 3 2 1 0 9 8 7 6 5 4 3 2 1 0 9 8 7 6 5 4 3 2 1 0

 INTEL RESERVED            V R   N IOP O D I T S Z   A   P   C
                           M F 0 T  L  F F F F F F 0 F 0 F 1 F
```

Flags in bits 0 to 15 are as for 80286. Bit 16 is used for resume flag and bit 17 for virtual mode flag.

Figure 3.13 *The EFLAGS register upper word.*

AX, BX, CX and DX registers can all be used in their 8-bit form as AH, AL, BH, BL and so on as for the previous generations of chips. Data can therefore be handled in 8, 16 and 32 bits in any one 32-bit register, and in 64-bit form by using two registers.

The segment selector registers are all 16-bit, but the older CS, DS, SS and ES set are supplemented by two more data registers FS and GS. The IP register is the lower word of the EIP 32-bit register and the FLAGS register is the lower word of the EFLAGS register. The EIP register holds the offset (relative to the address number in a segment register) for addressing, and the lower half of the register (IP) can be used for 16-bit addressing. In practice, this is all that is ever used to date because all of the software that is used in 386-based machines, other than by programmers, is 16-bit software written for the 8086. This is likely to change some day, but will be about as fast as the change to the metric system was (and still is).

The EFLAGS register consists of a lower word, FLAGS which is identical to the FLAGS register of the 80286 and is therefore used for 16-bit software written for the older chips. The upper word (see Figure 3.13), is mainly reserved, but contains as its lowest two bits the resume flag and the virtual mode flag. Of these the virtual mode flag can be set to enforce virtual 8086 mode while the 386 is running in protected mode. The flag can be set only in protected mode by an IRET instruction at highest privilege level or by task switching at any privilege level. Pushing the EFLAGS register on to the stack always resets the virtual mode flag, but the state of the flag will be preserved during task switching.

The resume flag is used along with debugging that uses breakpoints. Before a breakpoint is processed, the resume flag is checked and if the bit is set to 1, any debug fault will be ignored at the next instruction

CS-	32 bit Base Address	32 bit Segment limit	Segment attributes
SS-	32 bit Base Address	32 bit Segment limit	Segment attributes
DS-	32 bit Base Address	32 bit Segment limit	Segment attributes
ES-	32 bit Base Address	32 bit Segment limit	Segment attributes
FS-	32 bit Base Address	32 bit Segment limit	Segment attributes
GS-	32 bit Base Address	32 bit Segment limit	Segment attributes

Figure 3.14 *The set of descriptor (cache) registers, one for each segment register.*

and debugging will be resumed. The other flags, all in the lower word, perform the same functions as have been described for their 80286 equivalents.

There are also six 16-bit segment registers, each of which is associated with a descriptor (cache) register (Figure 3.14). This follows very closely the scheme that has already been described for the 80286, but using six registers rather than the four of the earlier chip. On the 386, there is no restriction on segment size, and the size of a segment can be as small as one byte or as large as the 2^{32} byte (4 Gb) limit of addressing with 32 address lines. The segment size will be fixed at 64K when the 386 is run in real mode so as to be compatible with a single 8086. The descriptor registers are used in protected mode to hold the usual base address, segment limit and access bits taken from the descriptor table, and in real mode will hold the base address for each memory reference (the code segment number shifted four-places left). As for the 80286, the descriptor register will be used to supply an address unless a change of segment has occurred. By using four separate data segments, the 386 chip allows for more use of these cache registers than was possible on the 80286.

The 386 uses three **control registers**, labelled as CR0, CR2 and CR3 (Figure 3.15). The designation of CR1 is reserved for use in future processors, though it has not been used in the i486. These registers are used to contain machine state information that is global as distinct from task state information, so that whatever is stored in these registers will affect all tasks that are running in the machine. Each CR register consists of 32 bits.

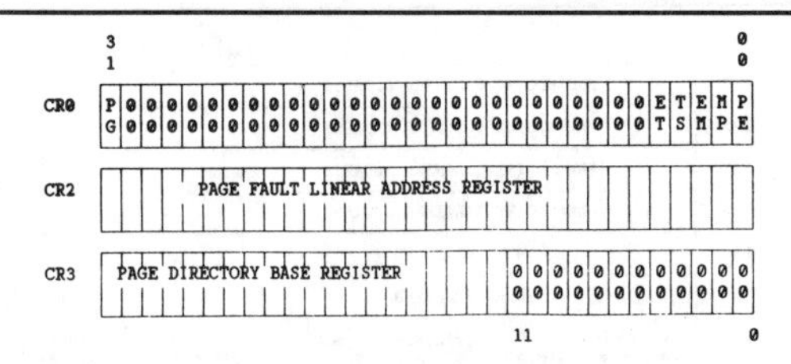

Figure 3.15 *The three control registers of the 80386DX.*

The CR0 register has its lower word identical in layout to the machine status word of the 80286. The LMSW instruction of the 386 will work in exactly the same way as on the 80286, and will ignore the upper word of the register. Software that is designed for the 386 and which does not need to be compatible with the older processor can use a different instruction (MOV CR0) to gain access to the entire CR0 register. This means that the protection enable bit cannot be reset by using LMSW, only by MOV CR0.

The main difference between 386 and 80286 lies in the use of bit 31 of CR0, the paging enable bit. Setting this bit enables the paging unit which is on the processor chip, and clearing this bit disables paging. The use of the paging unit will be described later in this section.

The CR2 register is used entirely as a page fault address register. When a page fault is detected, the address of the faulty page is put on to this register, and the page-fault handler stack contains an error code which provides information on the type of fault that was detected. The CR3 register uses 20 bits to hold the base address of the page directory table. Since the pages are of 4K, the lowest 12 bits of this address are ignored and will normally be zero.

Finally, the system address registers are familiar from 80286 practice. They consist of the descriptor registers (base address and limit) for the GDT (global descriptor table) and the IDT (interrupt descriptor table), along with the TSS (task state segment) and LDT (local descriptor table) selector registers. There is a descriptor cache register for the TSS and LDT selector registers of the form illustrated in Figure 3.14.

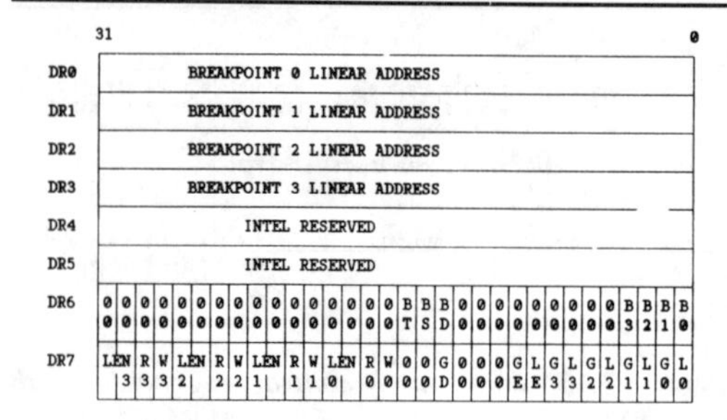

Figure 3.16 *The debug registers DR0–DR7 of the 80386.*

Unlike its predecessors, the 386 contains a comprehensive set of debug and test registers, all of 32 bits (Figure 3.16). Debug registers DR0 to DR3 contain addresses for up to four breakpoints, with DR4 and DR5 Intel reserved. DR6 stores the breakpoint status, and DR7 the breakpoint control. The status register holds the current state of the breakpoints, and the control register is used to set the breakpoints. If paging is not in use, the address held in the breakpoint registers will be identical to the physical address on the address lines. For a paged system, the address will be translated automatically.

The constitution of the control register is also shown in Figure 3.16. For each of the four breakpoints, the length field consists of 2 bits to specify the length of the breakpoint field as 1-byte (code 00), 2-bytes (code 01) or 4-bytes (code 11) so that the correct point for resumption can be found after the break. Note that code 10 is undefined and must not be used. If the break length is 2 bytes, then A_0 in the address is not used, and if the break length is 4 bytes, then neither A_0 nor A_1 will be used in the address.

The other two bits for each breakpoint are R and W respectively. If these bytes are 00, then any execution of the instruction will cause a break. For 01, only data write will cause activation, and code 11 will activate a break for reads or writes of data only. Once again, code 10 is undefined and must not be used. All of these breakpoint control bits are located in the upper word of the register.

The lower word of DR7 is used for access and

enabling bits. The set of G0, L0, G1, L1, etc. allows for breakpoint enabling to be global or local. If either a G or an L bit is set, then its breakpoint is enabled, subject to the correct address being present in the breakpoint register, and the correct setting of bits in the upper word of DR7. When a breakpoint is detected, by comparing the address in the breakpoint register with each instruction address in turn, then exception 1 is generated and the associated piece of code, the exception handler will attend to the requirements of the breakpoint.

When there is a task switchover, all of the **L** bits are reset to 0, but the **G** bits are unaffected. This prevents breakpoints that are local to one task from causing unwanted breaks in another task. There is **no** automatic restoration of these cleared breakpoints, so that restoration must be carried out by the software that is used for debugging. The **G** bits will cause breakpoints to be executed in all tasks, and for most purposes, since a user is unlikely to want the same breakpoints in each task, these will not be used.

The GE and LE bits are used for global or local exact breakpoint matching. An **instruction** breakpoint will **always** be reported exactly, but a **data** breakpoint may, because of pipelining, not be reported until several instructions later, perhaps not at all. If either **GE** or **LE** is set, then the normal action of the CPUs execution unit is altered to cause it to wait for the data operand to be transferred, so causing a break at exactly the point at which the data is read. The **LE** bit is always reset by a task switch (and can be reset only by software), but the **GE** bit remains set until it is reset by software.

The **GD** bit is used to provide an extra layer of protection. The use of a debugging program allows access to the chip registers and memory tables in a way that is most undesirable for an unauthorized user. The normal protection is that the debug registers are inaccessible except in real mode, or at privilege level 0 (the highest level) in protected mode. When the **GD** bit is set, **no** access is permitted to the debug register whatever, whether in real mode or at privilege level 0 in protected mode. When this bit is set, any instruction that attempts to use the debug registers will cause an exception 1 fault. The **GD** bit is cleared by the exception 1 handler.

The test registers of the 386 are TR6 and TR7, and

their use is rather specialized to the 386, so much so that their use is not guaranteed in future developments. The testing that they carry out relates to the translation lookaside buffer (TLB) which is used in paged mode (see later). Paged mode must be turned off when the TLB is to be tested in this way. The testing involves writing to the TLB memory and checking that a desired pattern can be found.

80386 modes and address formation

The 80386 can be used in real mode, in which state it operates as a fast 8086 machine with the ability to handle 32-bit numbers. Protected mode can be entered from real mode, and it allows the much more advanced memory management to be used. A variant of protected mode is virtual 8086 mode which alters the task switching so as to divide the memory into 1 Mb pieces each of which runs as if it were the memory space of an 8086, containing an operating system and an applications program of its own. These separate 8086 spaces will be protected from each other by the use of paging.

The 80386 can execute 16-bit instructions in both the real and the protected modes, using the D bit in the CS segment descriptor (see later). If this bit is 0, all operands and effective addresses are assumed to be of 16 bits as they would be for 8086 software. At the time of writing, this would be the normal state of use of the 386 chip, since no true 32-bit software is available. The setting of the D bit can be overruled if required by individual instructions that need to make use of memory outside the normal 8086 range.

Users of the 80386 refer to three types of address space as **logical**, **linear** or **physical**. Of these, the **physical** address is the number that appears on the 32 address pins of 80386. Physical address numbers can accommodate 4 Gb of memory, though at the time of writing few machines are provided with more than 4 Mb. The **linear** address is a 32-bit address which will exist in a register. For a non–page system, this will be identical to the physical address, but for a paged system the paging unit will translate the linear address into a different corresponding physical address.

The logical address (or **virtual** address) consists of a selector, which is a segment register content, and an offset obtained from the constituents of an address within a program. There are 16K possible segments and each offset can be of 4 Gb (2_{32}), making 64 Tb of this logical address space. The segmentation unit translates a logical address into a linear address, and the method that is used for this action differs according to whether real mode or protected mode is in use.

In real mode, the number contained in the segment register is simply left–shifted by four places and added to the offset in the way that is familiar in 8086 usage. This forms the linear address and also the physical address (no paging permitted), using only 20 of the 32 address lines. All segments are 64K long and start on 16-bit boundaries. An exception 13 is generated if an operand or instruction fetch causes the address to move past the end of a segment. The use of real mode is so perfectly compatible with 8086 addressing and with 80286 real mode that there is little point in describing it again.

In protected mode, the number contained in the segment register is used to obtain a descriptor table entry, and the base address in the table entry is added to the offset to form the linear address. This type of addressing has already been described in relation to the 80286. The diagram of Figure 3.17 is a reminder of how this system is used and of the sizes of quantities that are involved in addressing. Most of the machine-code instructions of the 80386 will imply the use of a specific segment, so that all code fetch actions use CS, operations of the PUSH and POP type make use of the stack segment, string move

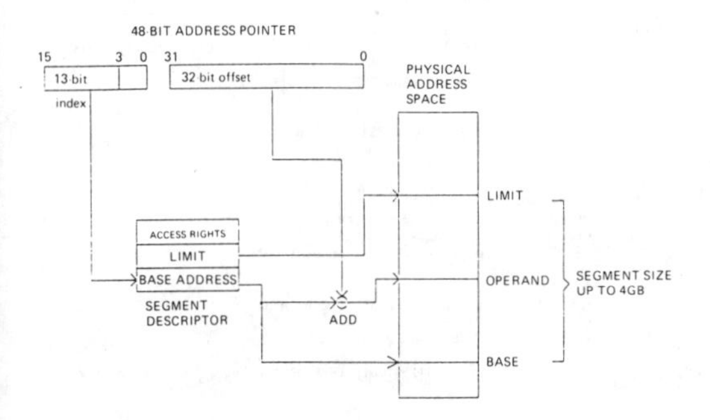

Figure 3.17 *How a linear address is formed in protected mode (ignoring paging).*

operations use ES and data references will generally use DS. For data references, the instructions provide for a segment prefix or override (see Section 8) which will override the automatic use of DS if there is a need to use any of the other segments.

The descriptor tables

The descriptor tables of the 80386 follow the pattern established in the 80286, so that this section will concentrate on the differences, with only an outline of the similarities. There are three table types, global, local and interrupt descriptor tables, each set up in RAM as arrays whose size can be as small as 8 bytes and as large as 64K. Each descriptor is of 8 bytes, so that a 64K descriptor table can hold 8192 descriptors. A segment register will use its upper 13 bits as an index into the descriptor table, using the descriptor table registers which hold the 32-bit base address and the 16-bit limit size for each of the tables.

The major difference between 80386 descriptors and 80286 descriptors lies in the use of the high-order word. For the 80286, this must be 0, but the 80386 in protected mode makes use of the arrangement illustrated in Figure 3.18, a segment descriptor, which shows the remaining bytes also. Concentrating on the most significant word, the lowest four bits are used to provide the most significant four bits of the segment limit number, making this a 20-bit

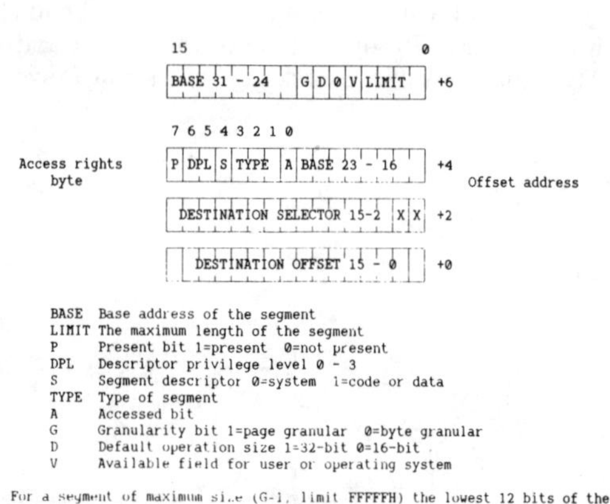

Figure 3.18 *The use of the high-order word of a segment descriptor in the 80386DX.*

number on the 80386 as compared to a 16-bit number on the 80286.

The AVL bit is a user-bit which is available for the operating system or for user software. The zero bit in position 21 **must** be zero to ensure compatibility with future processors. The D bit is used only for code segment descriptors, and defines the code as being 16-bit (D=0) or 32-bit (D=1).

The G bit, bit 23, is a **granularity** bit that shows how memory is divided up. If this bit is 0, then the segments can be of any size from 1 byte up to 1 Mb long only, but if this bit is 1 then the segments can be from 4K to 4 Gb long, in units of 4K (the page size). The granularity bit can be 0 or 1 irrespective of whether paging is enabled or disabled in protected mode. Bits 24 to 31 are used to carry the most significant 8 bits of the segment base address – note that the 32 bits of address are split into three locations in the descriptor bytes.

The system segment descriptors follow the general pattern as previously described, and the 4 bits allocated for type allow for the same selection as is permitted on the 80286 (refer back to Figure 3.8). Once again, the major difference between this and the 80286 are concentrated in the most significant word, which carries the highest bits of the base address and the limit address, with only the granularity bit available of the remainder.

The other descriptors of the 386 follow the pattern of the 80286 descriptors with two exceptions. In a call gate for the 386, the word count field is used to describe the number of 32-bit quantities copied from one stack to the other (the 80286 uses this number as the number of 16-*bit* quantities). The other point is the use of the **B** bit, which is bit 3 of the access byte in a call gate for the 80386, part of the type number bits. When this bit is 1, all PUSH and POP actions will be in 32-bit units; when the bit is 0 the 16-bit units are used.

Segments and cache registers

In protected mode, each segment register carries three data fields (Figure 3.19). Bit 2 is the table indicator so that with this bit zero, the global descriptor table is selected; with this bit equal to 1 the local descriptor table is selected. Bits 0 and 1 are used for the privilege level, allowing for levels 0 to 3 (in

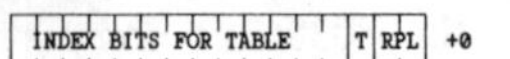


```
Bits 3 - 15 provide a 13-bit index number which can select up to 8192
descriptors.
     T=0 selects Global descriptor table
     T=1 selects Local descriptor table
     RPL 0 - 3 selects privilege level.
```

Figure 3.19 *The data fields of the segment registers in protected mode.*

descending order) of privilege. The remaining 13 bits form the index number for the descriptor, allowing 8192 descriptors, each of 8 bytes, to be obtained from one of the two tables that can be used, a total of 16K possible descriptors.

For each segment register (the name of selector register is better suited to the 80386) there will be a cache register which can hold the descriptor words. Each cache register will store the 32-bit base address, the 32-bit limit number, and a set of attribute bits from the descriptor. Note that these registers are much larger than the 48-bit cache registers of the 80286. The aim, however, is the same – to avoid the time penalty that attaches to the use of descriptors if they have to be fetched from memory for each addressing action. By using the cache registers, a full descriptor table read is used only when the content of a segment register is changed and the register then used. Task switching follows the same system as was outlined for the 80286.

Paging

Paging is a major area of difference between the 80386 and the earlier 80286 and 8088/8086 processors. The 8088/8086 CPUs divided memory into 64K segments, and this concept was followed also on the 80286. The 80386 allows for segments of virtually any size, and this created the need for another memory division system which used fixed sized units. The page is an arbitrary unit of 4K, and a page bears no relationship to any part of a program. The memory is often arranged so that a segment contains all of the code for a program, or all of its data and so on. A page will in general contain only a part of a program's code or data, but since many programs can spend a remarkable amount of time looping round a few instructions that are close to each other, a considerable amount of activity can be contained in the code that fits in a single page. Similarly, the 4K data page may contain all the data that is being accessed by the user over a considerable period of time.

Paging on the 80386 is a device used to translate a logical address into a physical address in protected mode. Paging has no application to real mode. The aims of using paging are that virtual memory can be used by moving pages between memory and disk and that a program can be run in different parts of the memory, by moving the position of its pages. Paging and segmentation are used together, with a page being a fixed-size piece of memory that will be a part of a segment. The size of page, 4K, has been fixed with the aim that it would correspond with the size of disk sector used on modern hard disks.

The paging scheme is illustrated in Figure 3.20. A linear address has been generated, and this is compared with previous values of linear address that reside in the translation lookaside buffer (TLB). When paging is put into use, this buffer contains various linear addresses and the corresponding physical addresses, and for some 98 per cent of address searches, the translated address will be in the buffer, directly usable.

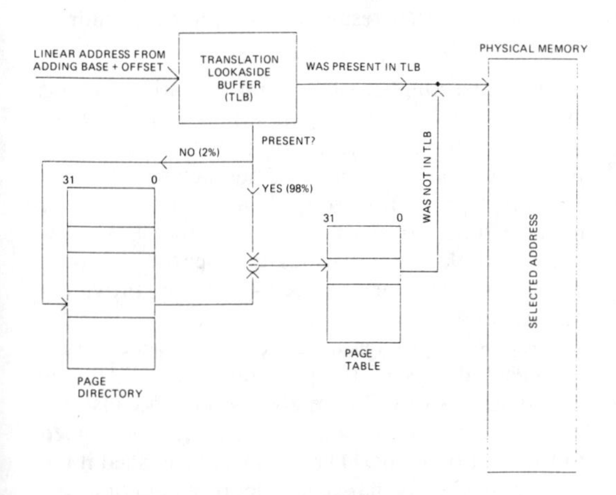

Figure 3.20 *Address paging, showing how a linear address is formed when paging is in use.*

If the linear address and its translation is not in the buffer, the translation must be made directly. This involves the use of two sets of tables, the page directory table and the page table. The linear address is used to provide three separate numbers that are used as location numbers in these tables. The lowest 12 bits are the **offset** which will be used to locate an address within the page (remember that the page is only 4K long). The next 10 bits are used to select a position in the page table. The upper 10 bits are used to select a position in the page directory. Both the page table and the page directory are 4K long.

An outline of the selection of an address is contained in the following steps, assuming that the address is in physical memory:

- Register CR3 stores the base address of the page directory. This address is added to the 10 directory bits of the linear address to point to a position in the page directory table.
- The entry in the page directory table provides the base address for the page table. This address is added to the 10 table bits in the linear address to point to a position in the page table.
- The entry in the page table points to a base address of a page in the main memory, and when the lowest 12 bits of the linear address are added, the result is the physical address number.

From the number of look-up operations in this cycle, it is obvious that paging would be too slow for practical use if it were not for the use of the translation lookaside buffer, combined with the fact that most programs use quite a restricted area of memory for a large part of their operations. We now need to look at the various table entries in more detail to see how they are used apart from providing address numbers.

The CR3 register is a 32-bit register which holds the base address of the page directory table. The lowest 12 bits of CR3 are always zero, because the page directory is itself a page, and it must be aligned to the start of a page. The TLB will be flushed if the number in CR3 is changed, either by a load or as the result of task switching.

The format of a page directory entry, which is

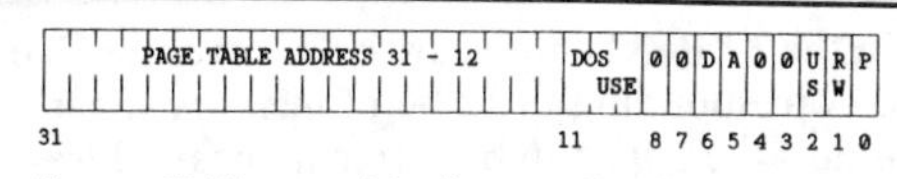


```
The upper 20 bits are used for the page table address
P=1   Use entry for address translation
P=0   Do not use entry for address translation
A=1   Access (read or write) to address is imminent
D=1   Write access is imminent
U/S and R/W are page level protection bits, see Fig.3.22
```

Figure 3.21 *The general format of a page directory entry.*

identical to that of a page table entry, is shown in Figure 3.21. The entry consists of 4 bytes, so that a 4K table contains 1024 entries, hence the use of only 10 bits from the linear address to find an entry – the 10 bits from the linear address are bits 2 to 11 in the pointer number since the lowest 2 bits must be zero to ensure that the pointer numbers move by 4 bytes when incremented. In the directory entry, the page table base address uses bits 12 to 31, a 20-bit address. Bits 9–11 can be used by the operating system and bits 8, 7, 4 and 3 must be zero. One suggested use for bits 9–11 is as a page–last–used code, so that when directory space is scarce the operating system can replace pages that have not been used for some time.

The **D** bit is used only in the page table entry, and is set to 1 when a write is to be made in the address range of the page. The **A** bit is used in both types of entry, and is set for any access, read or write. The **P** bit is used to distinguish an address in physical memory from one on a disk, so that when P = 1, the address can be translated into a physical address to be put on to the address bus, but when P = 0 the address is virtual. When P = 0, all of the other bits are available for OS use, so that up to 31 bits could be used to locate a disk sector.

The R/W and U/S bits are used as protection attributes, but distinguishing only two levels of privilege, supervisor and user. The supervisor level corresponds to privilege level 0, and user to levels 1, 2, or 3. Figure 3.22 shows the protection that is available from the use of these 2 bits.

U/S	R/W	Permitted to Level 3	Permitted to Levels 0,1,2
0	0	None	Read/Write
0	1	None	Read/Write
1	0	Read	Read/Write
1	1	Read/Write	Read/Write

Figure 3.22 *The use of the protection level bits.*

In the TLB, 32 entries are held, covering 32 pages and corresponding to 128K of memory. Each entry contains the upper 20 bits from a linear address and the corresponding page table base address, so that all that has to be done when a match is found is to add the offset bits from the linear address to the page table base address from the TLB.

We can now look in more detail at page selection, particularly at errors and virtual addresses. Assuming that the TLB cannot deal directly with a linear address, the page directory entry will be read, and if P=1 in the directory table and also in the page table entry, address formation proceeds as described earlier, providing a physical address.

If P=0 in the directory or in the page table, a page fault (exception 14) will be generated. This same exception will be generated if there is an attempted violation of protection, and in each case the linear address that caused the fault will be placed in register CR2. The register pair CS: EIP will hold the address of the instruction that caused the fault to appear, and the page–fault handler will push a 16-bit error code on to the stack. The least significant bits of this code are illustrated in Figure 3.23, and they show whether the error is due to a read or write action, the protection mode (user or supervisor) and whether the page is not in memory (P=0) or has failed because of protection violation (P=1). Note that Intel labels these bits in the same way as the bits in the table entries, but their message is different.

If the exception is generated because a page is not in memory, the operating system is responsible for dealing with the subsequent actions. This will require the OS to read the error-code zero order bit, and if the exception was caused by the page being absent, it must find a valid page on the disk, and swap this with a page in the memory, preferably one that has had

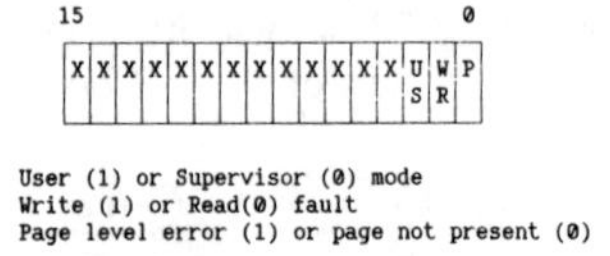

Figure 3.23 *The least significant bits of the error code.*

little recent use. The **A** and **D** bits in the table entries can be used by the OS to keep a count of accesses and writes to pages, so that a scale of relative use can be built up. Since all of this is an operating system function, it has no place in this book and we shall move on.

The virtual-8086 mode

Real mode allows the 80386 to be used as a fast 8086, with few advantages over the considerably cheaper 80286. Protected mode is of little interest at present to any user who wants to use industry-standard software. The virtual-8086 mode is therefore the aspect of 80386 use which ought to receive the most detailed attention, since it allows the 80386 to use multitasking along with existing software.

Virtual-8086 mode is a subset of protected mode, and can coexist with other uses of protected mode, so that one user can run Lotus 1–2–3 under MS–DOS and another can run whatever can be found to run under some version of Unix. Each user will be unaware of the other's use of the machine unless they are physically present in the same room trying to use the same keyboard.

The 80386 enters virtual-8086 mode when the VM bit, bit 17 in the EFLAGS register, is set to 1. This would normally be done in the course of a task switch action. The privilege level of any virtual-8086 task is always level 3, the lowest, because the use of the CS register for address formation in 8086 emulation precludes its use for holding privilege level bits as it does in normal use of protected mode.

The emulation of the 8086 implies the use of 20-bit addresses formed by left-shifting the contents of a segment register and adding the offset from the IP register (the 16-bit lower half of EIP). The total 1 Mb memory space corresponding to a 20-bit address amounts to 256 pages, and not all of the pages need to be in physical memory at the same time if the overall operating system can implement virtual addressing, though with 4 Gb of physical addressing available there would be no need for virtual addressing if sufficient RAM could be accommodated. The CR3 register which holds the base address for the page directory will be loaded each time the task is switched, so that each virtual task can use differently-located pages. In addition, however, it **is** possible for each task to have access to one common copy of the 8086 operating system, MS–DOS or DR–DOS usually, rather than use one copy in each virtual task.

The use of lowest privilege level for virtual-8086 mode makes several instructions (which in any case are not used in 8086 software) outside privilege level. Attempting to use such instructions in virtual-8086 mode will cause a type 13 exception. This is a major difference from real mode, because in real mode the privilege level is **automatically** set at 0, the highest level. The use of port addresses is covered by maintaining in the task state segment for each task an I/O permission bitmap, which details which I/O port numbers can be used in a particular virtual-8086 task.

The i486

The registers of the i486, and their uses, correspond very closely to those of the 80386DX, with the addition of the registers that are used in the floating-point unit and those used in conjunction with the cache memory. The floating-point registers follow the pattern that is described for the 80387 in Section 6. The cache test registers TR3, TR4 and TR5 (Figure 3.24), are used in the built-in self-test function of the chip, and play no part in the normal computing actions.

The TR3 cache test register controls access to the cache-fill buffer and cache-read buffer, and acts as a

NOTE: The main registers of the i486 consists of the register types that have previously been noted for the 80386DX and the 80387. In addition, registers are included for cache testing, as noted below.

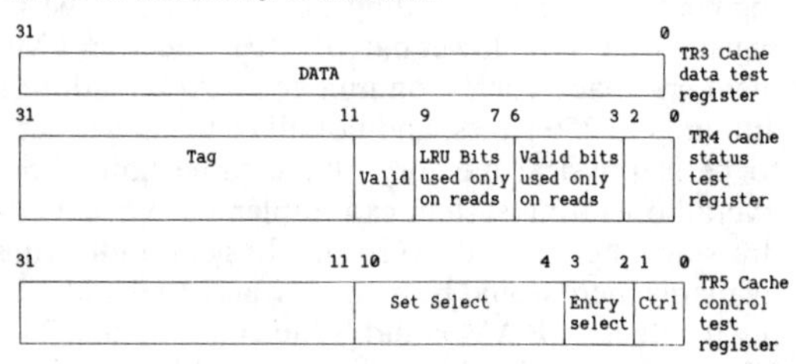

Figure 3.24 *The additional cache test registers of the i486. The other additional registers correspond to those of the 80387, see Section 6.*

Control bits Bit1	Bit0	Operation	Entry select bits Function	Set select bits
0	0	Enable W/R buffer	Select location	-
0	1	Perform cache write	Select entry	select write set
1	0	Perform cache read	Select entry	select read set
1	1	Perform cache flush	-	-

Figure 3.25 *The effect of the two control bits on read/write and cache actions.*

32-bit buffer for the 128-bit cache buffer, so that TR3 is used four times to fill or empty a cache buffer.

The TR4 register contains the cache tag, least-recently used (LRU) code and valid bit information during cache tests. This register must be loaded with a tag and valid bit before a write to a cache. When a cache has been read, the register will contain the tag and valid bit pertaining to that cache entry, along with the LRU bits and four valid bits from the data.

TR5 is the cache control test register which specifies which of the test operations will be performed and the set and entry (within the set) that will be read or written. Seven bits in this register are used to determine which of the 128 sets in the cache will be used, and the lowest two bits (Figure 3.25), select the actions of enable, write, read or flush.

NOTE: Initial values are shown for the two cases of the Built-in Self-test (BIST) function being activated or inactivated respectively.

Register	Initial value, with BIST	Initial value, no BIST
EAX	Zero (No fault indication)	Undefined
ECX	Undefined	Undefined
EDX	0004 + Revision I.D.	0004 + Revision I.D.
EBX	Undefined	Undefined
ESP	Undefined	Undefined
EBP	Undefined	Undefined
ESI	Undefined	Undefined
EDI	Undefined	Undefined
EFLAGS	00000002H	00000002H
EIP	0FFF0H	0FFF0H
ES	0000H	0000H
CS	F000H	F000H
SS	0000H	0000H
DS	0000H	0000H
FS	0000H	0000H
GS	0000H	0000H
IDTR	Base=0, Limit=3FFH	Base=0, Limit=3FFH
CR0	00000000H	00000000H
DR7	00000000H	00000000H
CW	037FH	Unchanged
SW	0000H	Unchanged
TW	FFFFH	Unchanged
FIP	00000000H	Unchanged
FEA	00000000H	Unchanged
FCS	0000H	Unchanged
FDS	0000H	Unchanged
FOP	000H	Unchanged
FSTACK	Undefined	Unchanged

Figure 3.26 *Initial register values of the i486, showing the effect of activating built-in self-test.*

Register values following a RESET are noted in Figure 3.26. The values for some registers depend on whether the built-in self-test (BIST) is activated or not, and if BIST has been used, the EAX register contains the results; a cleared register indicates that all tests give correct results. Following a RESET, instructions will be read starting from address FFFFFFF0H, but on the first intersegment jump or call the address lines A20–A31 will be forced LOW so as to execute in the lowest 1 Mb of memory. This allows the use of a ROM addressed at the highest part of physical memory to attend to initialization and resetting.

4 Processor signals

The deliberate similarities among the Intel chips imply that a description of signals for each type of chip would lead to considerable repetition. This section therefore follows the pattern of the preceding sections, examining the hardware signals of the 8088/8086 series first, and then looking at the differences that have been introduced with the later chips. The convention for writing inputs and outputs avoids the conventional bar to indicate active low, since these are not always obvious in diagrams or in text. All data and address buses are of the three-state type, so that specific mentions of three-state ability will be used only for control signal pins. Note the use of the # sign to mean active LOW.

The 8088/8086 processors

The simplest of the 16-bit processors is the 8088, and the main area of complication arises because of the dual use of pins, particularly comparing the MIN and the MAX configurations. In the following description, power supplies will be dealt with first, followed by non-data inputs (other than interrupts) and then outputs and input/output lines. Interrupts are dealt with last of all, because of the need for extended discussion of the software interrupt systems. Where an input action is described as **internally synchronized** this implies that a change of pin voltage may not take effect until one or more clock pulses later. Other inputs may be truly asynchronous, or may have to be supplied from circuits that synchronize the input to the clock. The bus timing will be dealt with in detail in Section 5, but references to the **bus cycle** mean the cycle of four clock pulses, T1 to T4 of the internal clock (four clock cycles).

The supply pins consist of pin 40 for the positive supply and pins 1 and 20 for earth (GND). The normal specification is $+5\,V \pm 10$ per cent with an

All timings in nanoseconds, ns.

Symbol	Timed period	8088		80C88A		80C88AL2		Notes
		Min	Max	Min	Max	Min	Max	
TCLCL	CLK cycle period	200	500	125	DC	125	DC	
TCLCH	CLK Low time	118		68		68		
TCHCL	CLK High time	69		44		44		
TCH1CH2	CLK rise time		10		10		10	1
TCL2CL2	CLK fall time		10		10		10	2

Note 1: From 1.0 V to 3.5 V levels
Note 2: From 3.5 V to 1.0 V levels

Figure 4.1 *The clock pulse specifications for the 8088 variants.*

absolute (transient) maximum rating of +7 V. A decoupling capacitor of around 10 nF should be connected between pin 40 and the nearest earth point at pin 1. Both GND pins should be earthed.

The CLK input at pin 19 requires a clock pulse with a duty cycle of 1/3. This will normally be provided from the Intel 82C84 chips (see Section 7). The maximum rise time is 10 ns, and the high and low times should be in accordance with the specification for the processor – Figure 4.1 shows the minimum high and low clock times for the various versions of the 8088. Note that bus timing diagrams can appear to be misleading, because they usually show a symmetrical clock wave-form.

The RESET input on pin 21, details of which follow later, is designed to restart the chip and is active HIGH. The TEST input on pin 23 is used by the software 'wait for test' instruction, and is active LOW, its normal state. In this state, the processor executes instructions normally, but when the TEST pin voltage is HIGH, the CPU will idle, waiting for the TEST voltage to drop again. The input is internally synchronized.

The READY signal input is on pin 22 and is active HIGH. The input must come from memory or a port to show that a data transfer can be completed. The input is not synchronized, and where the 82C84 clock generator is used this latter chip will provide for synchronization of the RDY signals from memory or ports. The set-up and hold time specifications of Figure 4.2 must be met if the READY signal operation is to be correct.

The MN/MX input on pin 33 is used to determine the mode of operation of the chip. With this pin voltage LOW, the 8088 operates in maximum mode, and the action of nine other pins is specific to this

Times	8088	80C88
READY Setup time (0 -1 transition) READY Hold time	118 ns 30 ns	68 ns after L/E of T2 20 ns

NOTES: L/E = leading edge. These are minimum times. Timings for 8086 processors follow the same pattern.
For READY to become inactive, the leading edge of the 1 to 0 transition should occur at least 8 ns before the clock trailing edge.

Figure 4.2 *Setup and hold times for the READY signal.*

mode which is designed to allow other devices (such as a co-processor) to take over the bus lines as and when needed. When the MN/MX pin voltage is HIGH the minimum mode is selected, in which the processor retains bus control at all times, altering the actions of the nine other pins referred to earlier. These pin actions will be dealt with separately.

Note that most of these input signals are required to be active HIGH, in contrast to the output control signals which are mainly active LOW. The only output signal which is unaffected by the MIN/MAX switching is the read strobe, RD, which is active LOW. This signal output goes LOW to indicate that a read of memory or I/O is being performed, making use of the local bus. The pin floats to third (isolated) state during a hold acknowledge – see Section 5 for signal timing.

The other outputs and inputs (other than interrupt) that are not affected by the MAX/MIN switch are the data and address buses, and these are divided into three groups. Address lines A8 to A15, on pins 2–8 and 39, are purely output lines, active HIGH. The address outputs from these lines remain valid through an entire bus cycle, and float to their isolated state only during an interrupt acknowledge or a local bus hold acknowledge.

The signals on pins 35–38 are output signals, but the use of these pins differs at different times in the bus cycle. In the first clock cycle, T1, these pins provide the four most significant bits of an address, active HIGH, and need to be latched. If a port is being used, the bits on these pins are all zero. During the other three phases of a bus cycle, these pins are used for status outputs irrespective of whether memory or a port was addressed. Pins 35–38 then provide signals S6–S3 respectively. The S6 output is always zero. S5 is the interrupt enable flag and S4/S3

S4	S3	Segment register
0	0	Alternate data (E)
0	1	Stack (S)
1	0	Code (C) or None
1	1	Data (D)

Figure 4.3 *The status S3–S4 signals and segment selection.*

are used to indicate which segment register is currently being used, following the pattern shown in Figure 4.3. In all bus phases, these lines will float to their isolated states during a local bus hold acknowledge.

The AD0–AD7 signals on pins 9–16 are for output of address bits, active HIGH, and the input or output of data bits. In the first clock cycle of the bus cycle, T1, these pins provide the low byte of an address (memory or I/O), which needs to be latched. In the remainder of the bus cycle, the pins serve for data input or output – remember that the data bus of the 8088 is only 8-bits wide. The lines float to the isolated state during a local bus hold acknowledge.

The MIN/MAX affected pins

The following pins are affected by the state of the MN/MX pin, and their different uses reflect the differing requirements of the chip in the two possible types of system. Typical systems are illustrated in Section 12. In the following descriptions, the nomenclature for pin function shows the MIN function first. Most of these pins are for outputs and many are active LOW.

Pin 24 is INTA (interrupt acknowledge) (QS1). In MIN mode this pin voltage goes LOW as a strobe output for interrupt acknowledge. The pin will be taken low in the T2–T4 clock phases in each interrupt acknowledge cycle. In MAX mode, the pin function is QS0, an output which along with the bit on pin 25 (QS1) indicates the status of the instruction queue of the 8088. The interpretation of these pins in MAX mode is illustrated in Figure 4.4.

QS1	QS0	Status
0	0	No operation
0	1	First byte of opcode from queue
1	0	Empty the queue
1	1	Subsequent byte from queue

Figure 4.4 *The instruction queue bits QS0 and QS1 in max mode.*

S2#	S1#	S0#	Status
0	0	0	Interrupt acknowledge
0	0	1	Read I/O port
0	1	0	Write I/O port
0	1	1	HALT
1	0	0	Code access
1	0	1	Read memory
1	1	0	Write memory
1	1	1	Passive

Figure 4.5 *The status S0–S2 signals in max mode.*

Pin 25 is ALE (address latch enable) (QS1). In MIN mode, this pin (active HIGH) has a HIGH voltage output during the first cycle T1 of the clock cycle, and is used to latch address bits (see above) for pins 9–16 and 35–38. There is no provision to float this output. In MAX mode, this pin provides the other queue status bit, QS1, noted above.

Pin 26 is DEN (data enable) (S0), an active LOW output. In MIN mode this can be used as an enable output for a data transceiver, since it is low for all memory and I/O address cycles and for interrupt acknowledge cycles. The pin is low during T2 and to the end of T4 in the cycle, but the point at which the voltage goes low differs. On a write cycle, the pin becomes active at the start of T2, but on a read or interrupt acknowledge cycle, the pin does not give its active signal until midway through the T2 clock cycle. The pin is floated during a local bus hold acknowledge. In MAX mode, this pin along with pins 27 and 28 is used as a status output, active LOW. The interpretation of the pin outputs in MAX mode is shown in Figure 4.5.

Pin 27 is DT/R (data transmit/receive) (S1), which is an output that in a minimum system provides either polarity of signal for a data transceiver, but is active LOW in a MAX system. In its MIN setting, a high output denotes transmit, a low setting receive. The pin voltage is valid from the last clock cycle (T4) **preceding** a read/write, and remains valid until the T4 cycle of the read/write. The pin voltage floats during a local bus hold acknowledge. In MAX mode, this pin provides the second status bit, S1, active LOW, see above.

Pin 28 is IO/M (input/output or memory) (S2), an output that determines whether data is transferred to or from memory or a port. In MIN mode, a low setting means a memory access and a high setting means port (I/O) access. The pin voltage is valid from the last clock cycle (T4) of the cycle **preceding** the data transfer and remains valid until the T4 cycle of the transfer. The pin is floated during a local bus hold acknowledge. In MAX mode, this is the third status pin, S2, active LOW, see above.

Pin 29 is WR (write) (LOCK), an output which is active LOW. In MIN mode this is the write strobe pin which is active in clock phases T2 and T3 (also during a wait) of the cycle. The write can be to memory or to a port depending on the setting of pin 28. The pin is floated during a local bus hold acknowledge. In MAX mode, the LOCK signal is active LOW and is an output that prevents any other chip from taking control of the bus. This signal can be activated by the LOCK prefix used on an instruction and will remain active until the next instruction has been completed, ensuring that an instruction will maintain use of the buses. The pin floats during hold acknowledge.

Pin 30 is HLDA (hold acknowledge) (RQ/GT1), used for output or input. In MIN mode, a HIGH on this pin is used to acknowledge a hold action, and at the same time the local bus and control lines will be floated to their isolated state. This occurs in the middle of the last clock cycle T4 of the clock cycle, or in an interrupt time. In MAX mode, the pin is one of a pair of request/grant pins with a lower priority than the other pin, pin 31. The pin is active LOW, with an internal pull-up resistor, and is used to control the use of the bus by another chip. A request/grant sequence is a form of handshaking which consists of a set of three pulses.

A pulse from another chip which holds the input LOW for one clock cycle (starting on any clock cycle) will initiate the request for the use of the local bus. Following this, on a T4 or a T1 clock cycle, the pin will be used as an output to acknowledge that the buses are floated and the **hold acknowledge** state will start at the next clock cycle. When the other chip is able to relinquish its use of the local bus, it sends a pulse which will hold this pin LOW for one clock cycle. The next clock pulse will be used as a T4 pulse (end of a bus cycle) and the CPU will regain use of the buses in this cycle.

The conditions for all of this are fairly stringent. There must be at least one clock cycle between each of the signals into or out of the RQ/GT pin(s). If the

IO/M#	DT/R#	SS0#	Status
0	0	0	Code access
0	0	1	Read memory
0	1	0	Write memory
0	1	1	Passive
1	0	0	Interrupt acknowledge
1	0	1	Read I/O port
1	1	0	Write I/O port
1	1	1	HALT

Figure 4.6 *Status signals in min mode.*

local bus was not in use at the time of a request pulse, the bus will be released on the next clock cycle or before the T3 portion of the next memory cycle. If the CPU is already performing a read/write action when the RQ pulse arrives, the bus will be released during the T4 clock of the bus cycle only if:

- the request is made before or during T2,
- the current bus cycle is not reading or writing the low byte of a word,
- the current cycle is not acknowledging an interrupt,
- the instruction is not a locked one (see LOCK).

Pin 31 is HOLD (RQ/GT0), structured as for pin 30. In MIN mode, this is an output which is taken HIGH to signal to another chip that the local buses are about to be held. The other chip will respond with the HLDA input (see pin 30), in the middle of the T4 clock (or during an interrupt). The HOLD output is not synchronized, and such synchronization should be added in order to ensure that the necessary set-up time (minimum 35 ns for the 8088 and 80C8AL, 20 ns for 8088–2, 80C88 and 80C88–2) is available. In MAX mode, the pin is used for bus request and grant signals as described for pin 30. This pin has priority over pin 30, so that if two other chips can issue bus requests, the chip connected to this pin will always have priority over the other chip.

Pin 34 is SS0, an output, active LOW. In MIN mode this is a status line equivalent to S0 in MAX mode. It can be used along with pins 27 and 28 to show the status of the CPU, as indicated in Figure 4.6 – note that this is very similar to the use of signals SS0–SS2 in MAX mode, but not identical. In MAX mode, pin 34 is **always** held high.

Interrupts

The hardware interrupt pins are the INTR pin 18 and the NMI pin 17. Of these two, the non-maskable interrupt (NMI) is provided as a last resort, a way of interrupting the action of the processor in an emergency such as a falling voltage on the supply line or a failure of memory. The INTR pin is used for maskable interrupts from external devices, and **internal interrupts** can also be generated. Because the interrupt mechanisms of the 8088 are so considerably enhanced compared to earlier 8-bit chips such as the 8080 and Z80 a full description of interrupt mechanisms is justified here, and most of it will be relevant also to the more advanced chips.

The NMI pin is normally low, and a non-maskable interrupt is generated when the voltage moves high – the interrupt is edge-triggered and need not be synchronized, though it must be present for at least two clock cycles. The NMI input signal must be a clean pulse which is free of bounces on the trailing edge. The current instruction will be completed and the current processor registers will be saved on the stack. The address of a routine for dealing with the interrupt is then obtained from a table in RAM (set up by the operating system). In the case of the NMI, this address is always the third in the table, interrupt type 2. The routine which deals with this interrupt has to be provided by the operating system, and will usually carry out a complete shutdown of processing, followed by a reset.

The INTR pin provides the alternative hardware method of interrupting the CPU from an external device. This input is level-triggered, active HIGH, and is sampled on the last clock cycle of each instruction, so that synchronization is internal. The action can be suspended by resetting the interrupt–enable bit in the flag register, so **masking** the interrupt. The arrival of an interrupt on this pin disables any further interrupts on this type.

Two complete bus cycles are used in an interrupt of this type. On the first, the LOCK output is forced LOW for four clock pulses (from T2 of the first cycle to T2 of the second), and one INTA output is delivered from the start of T2 to the end of T3. In all of this time, the AD0–AD7 bus will be floating. A second INTA occurs from the start of the T2 in the next cycle to the end of T3, and it is in this second bus cycle that the interrupt type number is transferred over the buses, which then revert to floating until the end of the interrupt.

When an interrupt signal is received on this pin, then, and is not masked, the external device that has caused the interrupt must also provide an interrupt number in the range 0 to 255. This 8-bit number will be read in through a port (one chip will normally provide both the port action and the interrupt signals) in the time of the second INTA signal, and used to locate an entry in the interrupt table, as above.

In this interrupt table, which is set up by the operating system, the interrupt numbers 0 to 31 are reserved as shown in Figure 4.7. Each table entry consists of 4 bytes, 2 words, and contains the CS and IP register values for a routine that will deal with the interrupt. Note that on the 8088 reserved interrupts beyond No. 4 are not used, but the 80386 uses numbers up to 16.

The interrupt numbers beyond 31 are used by the operating system and by other software. Provided that routines are written, table entries made, and devices programmed to deliver the correct interrupt number the user is free to make use of any interrupt numbers that have not been reserved by the operating system or for the chip.

Software interrupts or exceptions have the same

```
Int number          Reserved use
   00               Division underflow (divide by zero error)
   01               Single-step (for debugging)
   02               Non-maskable interrupt (NMI)
   03               Breakpoint (for debugging)
   04               Overflow in multiplication
Interrupt numbers 5 - 14H are available for the operating system, and the
following list indicates how these are used in PC machines.
   05               Print CGA screen
   06               Mouse button control
   07               Reserved
   08               System clock
   09               Keyboard
   0A               Real-time clock
   0B               Communications-1
   0C               Communications-2
   0D               Hard disk
   0E               Floppy disk
   0F               Printer
   10               VDU
   11               System configuration
   12               Memory size
   13               Disk input/output

Interrupts 14 - 1F should be considered as Intel reserved, but the operating
systems of many computers make use of these numbers to gain access to the BIOS
ROM routines.
```

Figure 4.7 *The arrangement of the interrupt table. The reserved numbers are not neccessarily honoured by manufacturers of computers, because the design of the 286 chip has become fixed and manufacturers do not feel obliged to provide for future versions.*

effect on the CPU as hardware interrupts delivered through the INTR pin. The INT n instruction in a program will cause an interrupt whose number is the 'n' of the instruction. For arithmetic actions, the INTO instruction can be used to provide for an interrupt occurring when a register overflows (that is, when the overflow flag becomes set). The IRET instruction is placed at the end of each software interrupt routine to signify a return from interrupt, so that registers can be reloaded from the stack and operation resumed.

The trap flag in the flags register of the 8088 will generate a type 1 interrupt at the end of each instruction. This allows for a routine to be used to pause the processor action, so that single-stepping can be carried out. A suitable routine will allow the user to display the contents of all registers and the stack at the end of each instruction. Such a routine is incorporated as part of the DEBUG diagnostic supplied with the MS–DOS operating system.

Port addressing

The 8088 can address up to 64K of single-byte ports or up to 32K of word-size ports (16-bit). The address lines A0–A15 are used for port addressing, with the upper 4 bits on lines A16–A19 always zero, and the M/IO output (pin 28) LOW to indicate port addressing. The difference between odd and even address numbers must be observed when connecting to ports – bits addressed by an even number are transferred on the D0–D7 lines.

Reset and initialization

The CPU is reset or started up by taking the RESET pin, pin 21, HIGH for more than four clock cycles. When the chip is first switched on, the voltage on the RESET pin must remain HIGH for at least 50 μs. The chip will end all actions on the leading edge of the RESET pulse, and remain inactive while the RESET pin voltage is held high. During this time, all three-state pins will be in their isolated state, ALE and HLDA are forced LOW, and the Status outputs are active for one clock period, then are floated. When the RESET pin voltage is taken LOW, an internal sequence takes up seven clock cycles, and following this time the 8088 will start to perform its memory read action at the fixed address FFFF0H.

The operating system must therefore use this address for the first instruction that the CPU is to act on, usually a jump to another address.

No use of the NMI signal can have any effect prior to the second clock cycle after the end of RESET, and in the remainder of the timed sequence that follows RESET, the NMI signal will not take effect immediately, but will act after a delay of one clock cycle. If a HOLD signal is applied immediately after a RESET, this will be put into effect after the delay and before the CPU has started its memory read at address FFFF0H.

HALT

The HALT instruction is issued by software, and is acknowledged by the CPU in a way that depends on the mode. In MIN mode, the ALE voltage goes HIGH one clock pulse after the execution of the instruction, and this signal can be latched. The status is then available continuously in the HALT state on the IO/M, DT/R and SSO pins. In MAX mode the HALT status appears on the S0, S1 and S2 pins, but the ALE signal will be obtained from the 8288 bus controller chip which will be present in a maximum mode system. A HALT is ended by a RESET or an interrupt request.

The 8086

The 8086 is so similar to the 8088 that it is easy to assume that the two are identical, but there are several important differences. The most obvious difference is the use of a 16-bit data bus, so that the address and data lines AD0–AD15 on pins 2–16 and 39 are all multiplexed. On the first clock cycle, T1, of a set of four, these pins are all used as output pins for the lower 16 bits of an address. For the remainder of a cycle in T2, T3 and T4 (and also during a wait cycle) the pins are input/output data pins. Latching must be used to separate the signals to the different buses. The upper 4 bits of any address are obtained by the multiplexed A16–A19 pins which are used also as status bits in the same way as was used on the 8088.

The bit A0 is used on the 8086 also as an enable output signal, so that any 8-bit device that needs to use the lower half of the data bus can be enabled by

BHE#	A0	Byte/Word selection
0	0	Whole word, even address number
0	1	Upper byte from/to odd address number
1	0	Lower bute from/to even address number
1	1	No effect

Figure 4.8 *The uses of the A0 and BHE signals in the T1 period.*

this signal in the first (T1) clock to transfer data in the remaining clocks of the cycle.

There are two significant changes in the use of the pins affected by the MIN/MAX setting. Pin 34 is no longer affected by the MIN/MAX setting, and pin 28 delivers an oppositely polarized signal output.

Pin 34 is now labelled as BHE/S7. On the first clock pulse of a bus cycle, T1, the bus high enable output goes LOW to signal that any 8-bit devices that need to transfer data to or from the upper byte (D8–D15) of the data bus in the remainder of the cycle (T2–T4) may do so. The combined effect of A0 and BHE in the T1 period is indicated in Figure 4.8. Note that this assumes the use of low bytes at even addresses and high bytes at odd addresses; the usual arrangement for 16-bit processors.

In the clock periods T2–T4 (and in T1 of a first interrupt cycle), the pin is used as a status-7 output, active LOW. This pin is spare, and the signal is not normally used.

Pin 28 is used in MIN mode for M/IO but with the important difference that the polarity of its output is reversed. Memory is addressed when this pin is HIGH, and a port is addressed when this pin voltage is LOW. This is the opposite of the way that this pin is used in the 8088. In MAX mode the pin provides a status signal as in the 8088.

The restart, halt and port information relating to the 8088 is all applicable to the 8086. The timed period following the HIGH to LOW change on the RESET pin takes about 10 clock pulses instead of the seven quoted for the 8088.

80188 and 80186

The 80188 and 80186 are internally equivalent to the 8088 and 8086 respectively, but with various peripheral chip actions included, replacing 15–20 chips which would be used externally in the corresponding 8088/8086 system. This affects the signals into and out from these chips because in many cases the signals concern the peripheral actions rather than the processor directly. The use of a 68-pin leadless chip carrier or the 68-pin pin-grid array or plastic leaded chip carrier package provides for the additional pins that are required for these signals. In the following description, only the signals that are peculiar to the 80188/80186 are dealt with in detail; signals that are identical to those of the 8088/8086 are noted without explanations.

The 80188

Supply pins are 9 and 43 for V_{cc} and 26 and 60 for earth (ground). All of these pins should be connected to their appropriate supply lines. There are three timing lines, two inputs and one output. The pins 58 and 59 are crystal inputs and the normal use of these pins will be to connect a suitable crystal across them, using the circuit illustrated previously in Figure 2.4. The crystal frequency should be double the desired clock frequency, so that for the 10 MHz version of the chip, a 20 MHz crystal (fundamental mode) should be used. As an alternative, an external clock pulse can be applied to pin 59 (X1). If this latter method is used, pin 58 (X2) should not be connected to any PCB track, in order to minimize stray capacitance.

Pin 56 supplies an external clock signal which has a 50 per cent duty cycle. There is enough drive current for direct connection to the 8087 arithmetic co-processor.

The RES input on pin 24 is designed to restart the chip and is active LOW. This state must be maintained for at least four clock cycles. When the RES pin goes low, all processing is terminated and all status lines are inactive for one clock cycle and then are floated. After at least four clock cycles if the RES pin is taken HIGH again, the processor will restart, and instructions will be fetched after about seven clock cycles. The action is internally synchronized. An internal Schmitt-trigger circuit is used to make it easier to ensure correct RES action by way of an RC time delay at switch-on.

The TEST input on pin 47 is used by the software 'wait for test' instruction, and is active LOW, its normal state. In this state, the processor executes instructions normally, but when the TEST pin voltage is HIGH, the CPU will idle, waiting for the TEST voltage to drop again. The input is internally synchronized. Interrupts will be serviced during a TEST wait.

The 80188 uses two forms of READY signal, ARDY and SRDY. The ARDY signal input on pin 55 is an asynchronous ready which is active HIGH to show that a data transfer can be completed. The leading edge of the pulse to this pin will be internally synchronized by the 80188, but the trailing edge must be externally synchronized to the system clock. If this input is not used, its pin should be connected to earth (ground). If the pin input is connected to V_{cc}, no wait states can be inserted.

The SRDY input on pin 49 is a synchronous ready input, active HIGH, whose input must be externally synchronized to the clock. If the pin voltage is V_{cc}, no wait states will be inserted. Either ARDY or SRDY must be used (either with a pulse or tied to V_{cc}) and if this input is not used it should be connected to earth (ground).

Timer input/output

Two timer inputs on pins 20 and 21 are used either as clock or as control signals, active HIGH. The form of signals that are required depend on how the programmable timer is being used, and the signals input to these pins will be internally synchronized. The corresponding outputs are on pins 22 and 23.

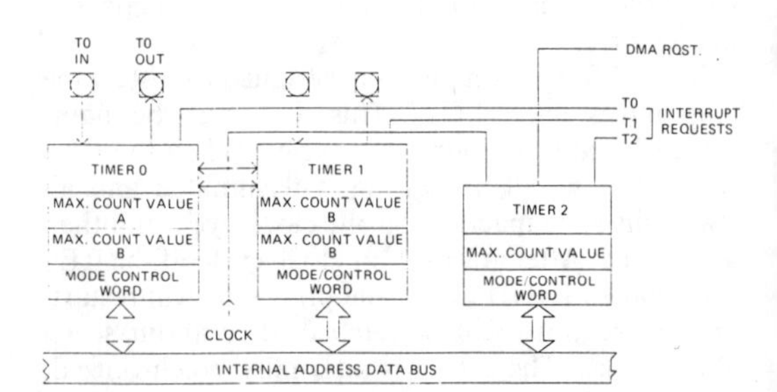

Figure 4.9 *The timer block diagram for the 80188/80186.*

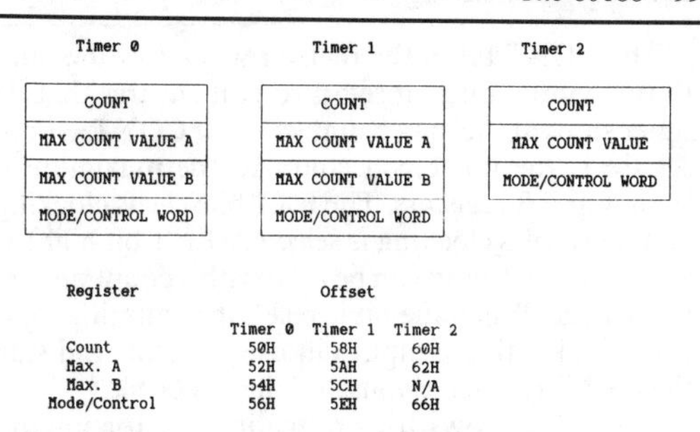

Register	Offset		
	Timer 0	Timer 1	Timer 2
Count	50H	58H	60H
Max. A	52H	5AH	62H
Max. B	54H	5CH	N/A
Mode/Control	56H	5EH	66H

Figure 4.10 *The registers in the 80188/80186 dedicated to timer use.*

EN	INH#	INT	RIU	0	....	MC	RTG	P	EXT	ALT	CONT
15	14	13	12	11		5	4	3	2	1	0

Note: Bits 6 - 10 are not used and should be masked in any test.

Figure 4.11 *The timer mode control register.*

The block diagram of the timer is illustrated in Figure 4.9, showing the use of the inputs and outputs for the two timers that are externally connected. The third timer provides an output to the other two, and can be used for time delays that are software controlled, as a pre-counter for the other timers, or as a source of DMA request signals.

Each timer is controlled by a set of registers, using four registers each for timer 0 and timer 1, and three registers for timer 2. These are 16-bit registers whose addresses in the internal peripheral control block are shown in Figure 4.10 – note that timer 2 (the internal timer) does not use a max count B register. The configuration of the timer mode/control register is shown in Figure 4.11.

The timer mode/control register is used as follows. If the ALT bit is 0, the max count A register will be used to decide the maximum count. If the ALT bit is 1, then the max A and max B registers will be used alternately, allowing the generation of asymmetrical waveforms. If the ALT pin is at level 0 the output pin will go low for one clock cycle after the maximum count is reached. If the ALT pin is high, the output pin will alternate in voltage as the different max count registers are used, 0 for B and 1 for A.

The CONT bit in the register will allow the timer to run continuously if set; if reset it ensures that the timer stops at the max count value. If CONT = 1 and ALT = 1, the timer will count to the maximum in both A and B registers. The EXT bit selects clocking, and if external clocking is selected (EXT bit high) an asynchronous input can be used with a count on each rising edge. When the bit is reset, the counting input is internal with the input pin used to stop and start the counting – see the use of the RTG bit.

The P bit allows for pre-scaling use for internal clocking only. With P = 0, the timer counting rate will be 0.25 × internal CPU clock rate. With P = 1, the output of timer 2 will be used as the clock input. The action of timer 2 will have to be arranged by the user. The RTG (retrigger) bit is used only if clocking is internal, and it controls the action of the input pin for the timer. For RTG = 0, timing operates for input high and is held while input is LOW. For RTG = 1, the first rising edge at the input pin will start the timer, clearing the count value initially. If CONT = 0 before the end of the count, then the timer will stop at the end of the count (resetting the EN bit, see following note), otherwise each rising edge on the input pin will result in a new count.

The EN bit enables control when set. With this bit zero no counting will take place and pulses at the input will be ignored. When CONT = 0, EN is cleared automatically at the maximum count value, preventing further counting. The INH bit allows EN to be updated. With INH high when the register is written, the EH bit will be written. With INH low during a write, the EN bit is **not** written. The INH bit is not stored in the register so that a read action **always** shows this bit low.

The INT bit, when set, enables interrupts from the timer at the end of each count. When the timer is using both A and B max count registers, an interrupt will be generated for each maximum count (in each register). The interrupt request is latched so that clearing this bit before the interrupt is serviced will not cancel the interrupt action. The MC bit is used to indicate max count by being set each time a max count is reached in either max count register. The bit has to be cleared by software; it does not reset at the next clock pulse following a maximum count. The RIU bit indicates which max count register is being used, with A = 0 and B = 1. This is a read-only bit.

On timer 2, the ALT, EXT, P, RTG and RIU bits are all permanently maintained at zero. On a reset, all timers will reset the EN and all selection bits, so that the timer out pins will go to high voltage on a reset.

DMA

Pins 18 and 19 are used for DMA request inputs, active HIGH. Pin 18 is DRQ0 and pin 19 is DRQ1, permitting two DMA channels to be used. Each DMA channel allows transfers of data from memory to memory or between I/O and memory without the intervention of the CPU. Each DMA channel makes use of a 20-bit pair of source and destination address pointers which can (optionally) be incremented or decremented after each transfer. Each transfer requires two bus cycles (8 clock cycles) so that the data transfer rate for an 8 MHz clock is 1 Mb/s. Figure 2.5 showed a block diagram for the DMA system.

Each of the DMA channels makes use of six registers in the internal control block, using the register addresses detailed in Figure 4.12. The source and destination pointers use two registers each, storing the upper 4 bits of each pointer separately from the lower 16 bits. The other two registers are the transfer count register and the control word register. The transfer count register is a 16-bit register whose contents are decremented after each DMA cycle. The source, count (if used) and destination pointer registers must be loaded before a transfer is started.

The arrangement of bits, and the functions of the bits in the control word register is shown in Figure 4.13. Synchronization can be switched as source,

Register	Address offset	
	Channel 0	Channel 1
Source pointer	C0H	D0H
Source upper 4	C2H	D2H
Dest. pointer	C4H	D4H
Dest. upper 4	C6H	D6H
Transfer count	C8H	D8H
Control word	CAH	DAH

Figure 4.12 *The registers, with addresses, for the DMA channels.*

M/ IO#	DEST DEC	DEST INC	M/ IO#	SOURCE DEC	SOURCE INC	TC	INT	SYN	SYN	P	TD RQ	X	CHG/ NOCHG#	ST/ STOP#	B#/ W
15	14	13	12	11	10	9	8	7	6	5	4	3	2	1	0

Figure 4.13 *The bits of the control word register.*

destination or none, and another bit determines whether or not an interrupt will be generated at the end of a transfer. The register can also be programmed so as to end transfer after a selected number of DMA cycles, and after each cycle the source and/or destination pointers can be decremented, incremented or allowed to remain constant. The relative priority of the channel (relative to the other channel) can be determined, and either source or destination pointer or both can be set to address memory or I/O ports. There are no restrictions on when the control word registers can be changed, but any change that is made during the time of a DMA transfer will affect that transfer.

Since DMA is independent of the CPU, transfers may be made with the CPU running or halted. This makes a difference only for destination-synchronized transfers, which are slower when the CPU is running because the DMA controller releases the bus after each transfer. A destination-synchronized transfer requires the destination of the data to initiate each transfer; a source–synchronized transfer required the source to initiate each transfer. Unsynchronized transfers take place continually until the correct count of words has been achieved. There is no specific acknowledge signal for a DMA transfer, and such a signal, if needed, will have to be obtained by using the chip-select lines.

DMA cycles always have priority over internal CPU cycles except when the DMA is between locked memory positions or when 16-bit word units have to be written to odd address numbers. The priority of one DMA channel relative to the other is determined by the setting of the **P** bits in the control word registers. When there is an external bus HOLD, this will have priority over a DMA action. The DMA action cannot be suspended by a maskable interrupt, only by the NMI, but a maskable interrupt received during DMA can affect the CPU after the DMA transfer has been completed.

Pin 50 is a HOLD input, active HIGH, which can be used asynchronously. When this pin is activated, the HLDA acknowledge signal will be available at pin 51 at the end of the T4 cycle (or idle cycle). At this time, the local buses and control lines will be floated. Lowering the level of the HOLD pin voltage will cause the 80188 to lower the HLDA voltage, and the buses will be released.

Address lines A8 to A15, on pins 1, 3, 5, 7, 10, 12, 14 and 16 are purely output lines, active HIGH. The address outputs from these lines remain valid through an entire bus cycle, and float to their isolated state only during an interrupt acknowledge or a local bus hold acknowledge, as for the 8088.

The signals on pins 65–68 are output signals, but the use of these pins differs at different times in the bus cycle. In the first clock cycle, T1, these pins provide the four most significant bits of an address, active HIGH, and need to be latched. If a port is being used, the bits on these pins are all zero. During the other three phases of a bus cycle, these pins are used for status outputs irrespective of whether memory or a port was addressed. Pins 65–68 then provide signals S6–S3 respectively. The S6 output is LOW to indicate a processor cycle, and HIGH to indicate a DMA cycle, but outputs S3, S4 and S5 are not used, and are held LOW in clock cycles T2 to T4. These lines will float to their isolated states during a local bus hold acknowledge.

The AD0–AD7 signals on pins 2, 4, 6, 8, 11, 12, 15 and 17 are for output of address bits, active HIGH, and follow the same pattern as the corresponding pins on the 8088.

All of the remaining pins of the 80188 are for output signals, many of them status or strobe outputs. The S7 output on pin 64 is held HIGH to indicate that an 8-bit data bus is being used – the 80186 uses this pin for a BHE signal, see later. Pin 61 is the ALE/QSO pin for address latch and queue status. At the trailing edge of this pulse, a valid address will exist on the bus, so that this edge can be used to latch addresses into the peripheral 8282 or 8283 chips. The pin is **never** floated. For queue status, see the description for pins 63 and 61.

Pin 63 is the WR/QS1 pin for write strobe and queue status. As a WR output, this pin will be active LOW during the T2, T3 and wait time of any write cycle; it will be floated during a HOLD. In queue status mode, this pin along with the output of pin 63 provides the information as detailed earlier in Figure 4.4.

Pin 62 is the RD/QSMD pin which is a read strobe or queue status mode pin. The voltage on this pin will go LOW in the T2, T3 and Twait cycles of a read action, and is guaranteed not to go LOW in T2 until the address bus has been floated. The pin floats during a HOLD, and is driven HIGH for one cycle during RESET. An internal pull-up will hold the pin voltage high if the pin is not driven by an external voltage. During RESET, the pin voltage is sampled to determine how pins 61 and 63 will be used. If this pin is (as is normal) HIGH, then pins 61 and 63 are used as ALE and WR respectively. If this pin is held LOW during a RESET, the pins 61 and 63 will be used to provide queue status as previously indicated.

The LOCK output is on pin 48, active LOW and activated at the start of the first cycle following the software LOCK instruction, for the duration of that cycle. As in the 8088, the LOCK output is used to ensure that other chips cannot take control of the buses. No data will be pipelined during a lock, and when several LOCK instructions are used in sequence there must be at least 6 bytes of code (which can be no-operation bytes) between the end of one LOCK and the start of another. The LOCK output is driven HIGH for one clock cycle during RESET and is then floated.

Pins 52 to 54 provide the S0 to S2 status outputs respectively. These are active LOW, and correspond exactly to the identically-labelled pins (26 to 28) of the 8088. Pin 40 is the data transmit/receive pin which controls the data flow direction through an external 8286 or 8287 chip. When this pin voltage is LOW, data is transferred to the 80188 and when this pin voltage is HIGH the 80188 will write data to the bus. An associated action is DEN on pin 39, active LOW. This is active during each memory or I/O access and can be used as an enable signal for the 8286 or 8287. The pin voltage is placed HIGH whenever there is a change of voltage at pin 40 (a change between reading and writing).

The remaining pins are selects for memory and peripheral chips. The 80188 allows memory to be organized (from the hardware point of view) in a way that is quite different from that used on the 8088. Three separate ranges of memory addresses can be selected, low, middle and high, with one select signal for low, one for high and four for the middle range. The starting addresses that can be used, and the memory size for each section can all be programmed by writing to the chip-select control registers (Figure 4.14), at addresses A0H to A8H. Memory size can be selected as 2K, 4K, 8K, 16K, 32K, 64K or 128K in

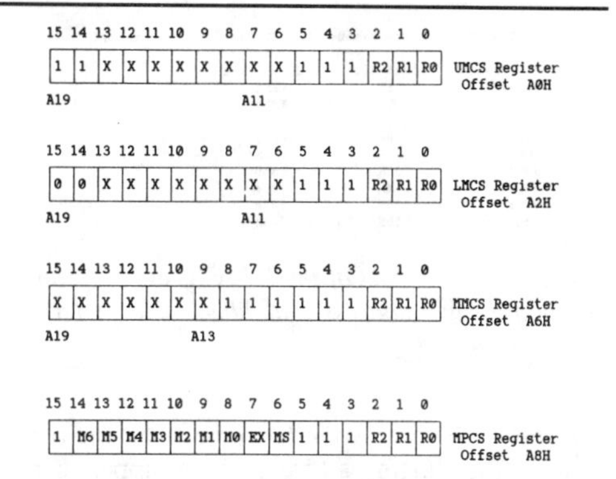

Note: Bits marked as X are used for block addressing, see text.

Figure 4.14 *The four chip-select control registers of 80188/80186.*

any of the ranges, and the lower and upper ranges also permit the use of 1K and 256K blocks.

The upper chip select (UCS), active LOW, is on pin 34 so that the voltage on this pin will go low whenever a software instruction requires the use of memory in the upper range. The upper limit of memory is always assumed to be FFFFFH, and the lower limit is determined by the value stored in the UMCS register. This register uses bits 6 to 13 to specify the lower memory limit for the upper memory range, with bits 0, 1 and 2 used to specify WAIT states and the response to the RDY signal (Figure 4.15). Bits 3, 4, 5, 14 and 15 must all be '1' bits. When a memory reference is made, the highest 10 bits of the address are compared to the highest 10 bits of the UMCS register and if the address bits correspond to a number that is equal to or higher than the UMCS number, then the UCS signal is activated. Figure

R2	R1	R0	Number of wait states.
0	0	0	No wait states, external RDY used
0	0	1	1 wait state, external RDY used
0	1	0	2 wait states, external RDY used
0	1	1	3 wait states, external RDY used
1	0	0	No wait states, external RDY ignored
1	0	1	1 wait state, external RDY ignored
1	1	0	2 wait states, external RDY ignored
1	1	1	3 wait states, external RDY ignored

Figure 4.15 *The coding of bits 0–3 in the UMCS register.*

NOTE: The UMCS numbers assume that R2=R1=R0=0. All numbers in hex.

Base Address	Block Size	UMCS Value
FFC00	1K	FFF8
FF800	2K	FFB8
FF000	4K	FF38
FE000	8K	FE38
FC000	16K	FC38
F8000	32K	F838
F0000	64K	F038
E0000	128K	E038
C0000	256K	C038

Figure 4.16 *The bits of the UMCS register, showing the coding base address and block size in terms of UMCS value.*

4.16 shows the correspondence between the register settings, the starting address of the upper memory block and the upper memory size. This table assumes that the three selection bits in the register are all at zero – their values will be assumed to be zero in memory comparisons because only 10 bits are compared.

After a RESET, the register is set for a 1K memory block and will need to be rewritten if a larger area is needed. Any number placed into the select register that does **not** correspond with the numbers in Figure 4.16 will cause undefined operation, so that care is needed in designing the software that will load this and the other memory-select registers. After a RESET, the 80188 will start executing instructions stored starting at address FFFF0H, so that the default selection of a 1K block in this range is appropriate.

The lower memory chip select (LCS), is on pin 33, active LOW. This pin becomes active whenever a memory address in the lower area of memory is to be used, and is not floated during a bus HOLD. Like the upper address range, this area is selectable by the use of a register, the LMCS register at address offset A2H. The layout of this register is as for the UCS register but with bits 14 and 15 always '0'. Bits 0, 1 and 2 are used as selectors for WAIT states and RDY response as before.

The start of memory is at address 00000H, with the interrupt vector table starting at this address. The upper limit of low memory is defined by the number inserted into the bits 6 to 13 of the LMCS register, using the same convention of comparing the top 10 bits of the register with the top 10 bits of the (20-bit) memory address. The correspondence between the

NOTE: The LMCS numbers assume that R2=R1=R0=0. All numbers in hex.

Upper Address	Block Size	LMCS Value
003FF	1K	0038
007FF	2K	0078
00FFF	4K	00F8
01FFF	8K	01F8
03FFF	16K	03F8
07FFF	32K	07F8
0FFFF	64K	0FF8
1FFFF	128K	1FF8
3FFFF	256K	3FF8

Figure 4.17 *The bits of the LMCS register, showing the coding of upper address and block size.*

number in the LMCS register (assuming that bits 0, 1 and 2 are all zero) and the memory limits are illustrated in Figure 4.17. Any other numbers used in this register will cause undefined actions. A RESET will leave this register undefined, and the corresponding chip select pin (33) will **not** be used until the register has been written.

The mid-range memory select pins are 35 to 38, providing MCS3 to MCS0 respectively, all active LOW. Like the other selects of this kind, these pins are not floated during a bus HOLD. Any part of the total of 1 Mb of selectable memory that has not been defined as upper or lower can be selected by using these signals, and the memory that is allocated as mid-range memory will automatically be divided up among the four chip-select signals. For example, if the mid-range memory consists of 64K, then the memory will be selected in 16K units, each 16K by a different select signal.

The allocation of memory is determined by the settings in the MPCS register at offset address A8H (Figure 4.18). Bits 8 to 14 (labelled as M0 to M6) are used, and **only one of these bits must be set**. If more than one of these bits is '1', the results are undefined. Figure 4.18 shows the settings of bits 14 to 8 and the

NOTE: The MPCS numbers assume that R2=R1=R0=0. All numbers in hex.

Total Block Size	Select Size	MPCS Bits 14-8
8K	2K	000
16K	4K	002
32K	8K	004
64K	16K	008
128K	32K	010
256K	64K	020
512K	128K	040

Figure 4.18 *The bits of the MPCS register, showing the coding of block size and select size.*

corresponding memory block size and the amount of memory handled by each select signal. In each case, MCS0 handles the lowest unit of memory and MCS3 the highest.

The MPCS register handles block sizes, but the base address of the mid-range memory is determined by the settings of the MMCS register (see Figure 4.14). Bits 3 to 8 of this register are always set, with bits 9 to 15 used to determine mid-memory base address. These bits correspond to bits A13 to A19 in an address, with bits A0 to A12 assumed zero. The lowest possible memory starting address is 02000H, which corresponds to a location which is 8K from the start of memory.

The starting address in the MMCS register **must** be a multiple of the memory block size, so that for an 8K block (each select handles 2K) a starting address of 8K (which is 02000H), 16K (04000H) and so on could be used (00000H could also be used if no lower memory space were being separately defined). The highest possible start-address setting is FE000H, but this is not likely to be used in practice.

The contents of both MPCS and MMCS registers are undefined after a RESET, so that none of the mid-range select pins will be active until the registers have been written. Particular care needs to be used to ensure that the starting address put into the MMCS register is an exact multiple of the block size placed in the MPCS register.

The remaining pins are used for peripheral chip select signals PCS0 to PCS6. Of these, pins 25, 27, 28, 29 and 30 are used for the PCS0 to PCS4 signals respectively. Pin 31 is used for the dual PCS5/A1 signals and pin 32 for the dual PCS6/A2 signals. The A1 and A2 signals are latched address bits which can be used to provide A0 and A1 address bits for 8-bit peripheral chips, and the switch between peripheral chip select and peripheral chip address for these two pins is done by means of a bit in the MPCS register, see later.

Each of the peripheral chip select signals is active LOW when an instruction requires a port address in the normal 64K port address range or if a peripheral needs to use an address in the normal memory range. Like the memory select outputs, these lines are not floated during a bus HOLD. The MS and EX bits in the MPCS register are used as indicated in Figure

Bit	Action
MS=1	Peripherals mapped into memory space
MS=0	Peripherals mapped into I/O space
EX=0	5 PCS# lines, A1, A2 provided
EX=1	7 PCS# lines, A1, A2 not provided

Figure 4.19 *MS and EX bits and their control effects.*

15	14	13	12	11	10	9	8	7	6	5	4	3	2	1	0	
X	X	X	X	X	X	X	X	X	X	1	1	1	R2	R1	R0	PACS Register Offset A4H
A19									A10							

Figure 4.20 *The PACS register bits in detail.*

NOTE: PBA = Peripheral base address

PCS Line	Range of locations
PCS0	PBA to PBA+127
PCS1	PBA+128 to PBA+255
PCS2	PBA+256 to PBA+383
PCS3	PBA+384 to PBA+511
PCS4	PBA+512 to PBA+639
PCS5	PBA+640 to PBA+767
PCS6	PBA+768 to PBA+895

Figure 4.21 *The memory areas handled by the PACS register.*

4.19 to determine whether the peripherals are mapped into memory space or I/O space, and whether 5 or 7 PCS lines will be used – if 5 PCS lines are used, then pins 31 and 32 are used for the A1/A2 signals which will select internal registers in 8-bit peripheral chips.

The peripherals are assumed to be addressed in 128-byte intervals for up to 5 or 7 consecutive blocks whose starting address is determined by the number in the PACS register at offset A4H (Figure 4.20). In this register, bits 6 to 15, corresponding to address pins A10 to A19, are programmable, with bits 3 to 5 all set, and bits R0 to R2 used in the usual way to specify RDY mode for PCS0 to PCS3 only. If the chip-select block of addresses is located in I/O space (so that port addresses are being used), then bits 12–15 in the PACS register **must** be zeros because the I/O address range is only 64K and cannot use the address lines A16–A19. The memory area handled by each peripheral chip select line, relative to the base address is illustrated in Figure 4.21. As before, both PACS and MPCS registers are undefined after a RESET, and the pin outputs will not be available until the registers have been written.

Interrupts
NMI for the 80188 is on pin 46, internally synchronized, and in every respect identical to the NMI of the 8088/8086. The 80188, however, used four maskable interrupts INT0–INT3. INT0 (pin 45) and INT1 (pin 44) are interrupt inputs, and INT2 (pin 42) and INT3 (pin 41) can also be used as interrupt–acknowledge outputs. When these pins are used as inputs, they are active HIGH, and when INT2 and/or INT3 are configured (by software) to be acknowledge outputs they are active LOW. All of the inputs can be software-selected to be edge triggered or level-triggered. The action of all four pins changes when the interrupt controller section of the chip is used to select slave mode.

The 80186

The differences between the 80186 and the 80188 are as small as the differences between the 8086 and the 8088, and are mainly concerned with the use of 16 data lines, multiplexed with the address lines, on the 80186. Pins 1–8 and 10–17 are all AD lines (AD0 to AD15) on the 80186, and pin 64 is a dual BHE/S7 output.

In the T1 part of a bus cycle for read, write or interrupt cycles, pin 64 provides the BHE (bus high enable) signal, active LOW, to allow data to be enabled on to the high-order pins D8–D15 of the data bus. During T2 to T4 parts of the cycle, pin 64 provides status information in combination with A0, using the same scheme as for the 8086 (described earlier).

Other differences are in internal architecture and these affect execution times rather than hardware arrangements. The queue length in the 80186 is 6 bytes (3 words) rather than the 4 bytes of the 80188. The 80186 will not fetch an instruction into the queue until there is a 2-byte space; the 80188 will fetch when a single byte space appears. The 80186 is also faster to execute 16-bit fetches and writes.

The 80286

By comparison with the 80188/80186 chips, the 80286 looks considerably simpler in hardware terms.

The same 68-pin packages are used as for the 80188/80186 series, and many actions are sufficiently similar to those of the 8088/8086 or 80188/80186, so only a brief reminder is given here.

DC requirements are for pins 30 and 62 to be used for V_{cc} and pins 9, 35 and 60 for V_{ss} (earth or ground). All of these pins **must** be connected to their appropriate supply lines. In addition, a substrate filter capacitor of 47 nF **must** be connected from pin 52 to earth (ground). The maximum allowable leakage current for this capacitor is 1 μA. This capacitor will charge for a maximum time of 5 ms after power and clock pulses have been applied, and RESET should be used during this time. Charging takes place by way of the substrate bias generator, and the capacitor ensures that the substrate voltages remain below V_{cc} during this initial period. After the capacitor has charged, the clock of the 80286 can be synchronized to another system clock by pulsing RESET low at a system clock pulse.

Like the 8088/8086, the 80286 relies on an external clock generator, and the input from this chip is connected to the CLK pin, pin 31. The clock frequency is internally divided by 2 for the processor clock, and the division can be synchronized to an external clock as described above.

The RESET input on pin 29 is active HIGH, and it has the effect of clearing any internal settings. The HIGH state must be maintained for at least 16 clock cycles. The action is internally synchronized, and external synchronization is needed only if the processor clock must be synchronized to another clock. If external synchronization is used, the processor cycle will end at the second falling edge of the system clock. During the HIGH period of the RESET pin, the signals S0, S1, PEACK, A0–A23, BHE and LOCK are all high, signals M/IO, COD/INTA and HLDA are all LOW (but HLDA is low only if HOLD is also LOW), and the data bus pins are floated.

The 80286 resumes processor action when the RESET voltage goes LOW, an action which **must** be synchronized to the processor clock. The first 38 clock cycles will then be used for initialization before the first byte fetch cycle of an instruction can start.

The READY signal input is on pin 63 and is active LOW, at which voltage bus operation is normal.

With READY HIGH, the processor will wait for as long as is needed until the READY signal goes LOW again. READY is ignored during a bus-hold acknowledge (pin 65).

The HOLD (pin 64) and HLDA (pin 65) signals are used in allocating bus control. A HOLD input, active HIGH allows another chip to take command of the buses, and the response of the 80286 will be to float the bus lines and to activate HLDA (active HIGH). This remains in force until the external chip lowers the voltage of the HOLD pin again, which will result in the HLDA voltage dropping and the buses being driven again from the 80286. The HOLD input can be asynchronous.

Four pins are devoted to the use of a processor extension chip (co-processor). The co-processor interface uses I/O port addresses 00F8H, 00FAH and 00FCH which are in the Intel-reserved set of port addresses. The PEREQ input on pin 61 is active HIGH, and requests the 8086 to transfer data to a co-processor. This pin will float during a bus-hold acknowledgement. The PEACK signal, active LOW, on pin 6 is used to signal to the co-processor when the requested data is being transferred. The signal from PEACK can be asynchronous.

The BUSY input to pin 54, active LOW, stops program execution on a WAIT and also on some ESC instructions until the input is put HIGH again. During this form of wait, an interrupt can be received if necessary. The ERROR input on pin 53, active LOW, will cause the 80286 to be interrupted during a WAIT or some ESC commands, and to be serviced by a co-processor routine. Both of these inputs use internal pull-up resistors and can be asynchronous.

The use of the 68-pin package allows a full set of data and address pins, all active HIGH, to be used. Pins 36 to 51 inclusive are used for data, with the bits of the lower half of the data bus interleaved with the pins of the upper half in the sequence D0, D8, D1, D9, D2, D10 and so on. The data bus floats during bus-hold acknowledge. The address pins are 7, 8, 10 to 28 and 32 to 34, a total of 24 pins. The value of A0 is low for transfers using D0–D7 so that the lower byte of a word is on an even-numbered address. The upper pins A16 to A23 are all held at zero during a port transfer, since only 64K of port addresses can be used. The address bus floats during a bus-hold acknowledge.

COD/INTA#	M/IO#	S1#	S0#	Form of Bus Cycle
0	0	0	0	Interrupt acknowledge
0	0	0	1	Forbidden
0	0	1	0	Forbidden
0	0	1	1	No meaning
0	1	0	0	HALT if A1=1, else shutdown
0	1	0	1	Memory data read
0	1	1	0	Memory data write
0	1	1	1	No meaning
1	0	0	0	Forbidden
1	0	0	1	I/O port read
1	0	1	0	I/O port write
1	0	1	1	No meaning
1	1	0	0	Forbidden
1	1	0	1	Memory instruction read
1	1	1	0	Forbidden
1	1	1	1	No meaning

NOTE: States listed as forbidden will not occur. Those listed as no meaning may occur but not in a normal bus cycle.

Figure 4.22 *The status signals of the 80286.*

The bus-high enable output is on pin 1, active LOW, and becomes active during transfer along D8–D15. This and the A0 voltage can be used to indicate data status, as on the 8086 (see Figure 4.8).

Pins 5 and 4 are S0 and S1 respectively, active LOW, floating during a bus-hold acknowledgement. These pins are used rather as the corresponding pins of the 8086, but in conjunction with the M/IO and COD/INTA pins (67 and 68 respectively) to indicate the status of a bus cycle. The table of Figure 4.22 shows how combinations of these signals are interpreted.

The M/IO output signal on pin 67 is used HIGH to indicate memory access and LOW to indicate I/O port access. The pin floats during a bus-hold acknowledgement. The COD/INTA output signal on pin 68 is HIGH on an instruction fetch cycle and LOW for a memory data read cycle. Its use along with M/IO and the status pins allows an interrupt acknowledge cycle to be distinguished from an I/O cycle. The pin floats on a bus-hold acknowledgement.

The LOCK output, pin 68, is active LOW to indicate that other chips must not gain the use of the buses. The LOCK output can be activated by the LOCK software instruction, or automatically during memory exchange instructions (the XCHG command), interrupt acknowledge or descriptor table access.

Interrupts

The hardware interrupt pins are the INTR pin 57 and the NMI pin 59, and the general remarks as for the 8088 interrupt system apply also to these inputs.

INT No.	Action	Related Opcode	Notes
0	Divide error	DIV, IDIV	1
1	Single step	Any	
2	NMI	NMI pin	
3	Breakpoint		
4	Overflow	INTO	2
5	BOUND range exceeded	BOUND	1
6	Invalid opcode	Any incorrect	1
7	No co-processor	ESC, WAIT	1
8-15	Intel reserved		3
16	Co-processor interrupt	ESC, WAIT	
17-31	Intel reserved		3
32-255	User defined.		

NOTES: 1 - Return address points to the instruction that caused interrupt
2 - Return address does not point to instruction that caused interrupt
3 - Do not use

Figure 4.23 *Software interrupt numbers for the 80286.*

The NMI pin is active HIGH, and is used asynchronously. The pin voltage must have been normal (LOW) for at least four clock cycles before the interrupt, and needs to be kept high for at least four clock cycles to initiate the interrupt correctly. As for the 8088/8086, the NMI interrupt uses the vector 2 in the interrupt table.

The INTR pin (pin 57) provides the alternative hardware method of interrupting the CPU from an external device, and the action is identical to that for the INTR pin of the 8088/8086. Figure 4.23 shows the Intel recommended uses of software interrupts and exceptions, using the INT n software instruction.

Port addressing

The 80286 can address up to 64K of single-byte ports or up to 32K of word-size ports (16-bit). The address lines A0–A15 are used for port addressing, with the upper 8 bits on lines A16–A24 always zero, and the M/IO output (pin 67) LOW to indicate port addressing. The difference between odd and even address numbers must be observed when connecting to ports – bits addressed by an even number are transferred on the D0–D7 lines, so that devices which use 8-bit buses can be connected either to the lower 8 data lines (data on even address numbers) or on the upper data lines (data on odd address numbers). For better control, the 8259A interrupt controller chip should be connected to the lower data byte D0–D7, to ensure that the correct vector is returned.

```
NOTE: All numbers in Hex

Register        Value after RESET
  Flags              0002
  MSW                FFF0
  IP                 FFF0
  CS                 F000
  DS                 0000
  ES                 0000
  SS                 0000
```

Figure 4.24 *States of registers following a RESET.*

Reset and initialization

The CPU is reset or started up by taking the RESET pin, pin 29, HIGH for more than 16 clock cycles. When the chip is first switched on, the voltage on the RESET pin must remain HIGH for at least 5 ms (see CAP pin, pin 52). The chip will end all actions on the leading edge of the RESET pulse, and remain inactive while the RESET pin voltage is held high. During this time, all three-state pins will be in their isolated state. When the RESET pin voltage is taken LOW, an internal sequence takes up to 32 clock cycles, and following this time the 80286 will start to perform its memory read action at the fixed address FFFF0H. The operating system must therefore use this address for the first instruction that the CPU is to act on, usually a jump to another address. The 80286 will be in REAL mode at the time of reading its first instruction. In addition, some register values will be set as indicated in Figure 4.24. The HOLD pin must not be made active during a RESET, nor until 34 clock pulses after the trailing edge of the RESET signal.

The 80386DX and /SX chips

The 80386DX and 80386SX chips can be considered together, because the only significant hardware signal differences are the number of address and data bus lines. Note, however, that the chips use very different packages, so that they are not interchangeable. The use of the large number of V_{cc} and V_{ss} connections was noted earlier. A minor point is that Intel data sheets for these chips use the convention of marking active LOW signals with a hashmark, as used throughout this book. The pins are also located by using a number/letter grid on the 80386DX.

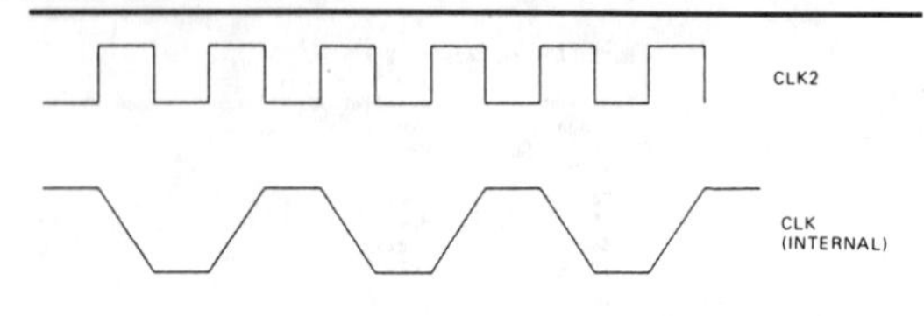

Figure 4.25 *The CLK2 and internal clock phase signals.*

The CLK2 input is on pin F12 of the 80386DX and pin 15 of the 80386SX. As usual, the external signal into this pin is divided by 2 to form the CPU clock, and the relationship between this pin input and the internal clocks are shown in Figure 4.25 for the 20 MHz version of the 80386DX chip. Other clock information is as for the 8086 and 80286. The CLK2 period is used extensively in data sheets as a unit of time.

The NA input on pin D13 of the 80386DX, pin 6 of the 80386SX is active LOW and a change in value must be synchronized to the processor clock. This is the next address request signal that is used in address pipelining. When this pin is put low, the processor will output the next address on to the address bus if the next bus request is already available. The NA pin input can be tied to a steady voltage, high for non-pipelined operation or low for pipelined operation, or the value can be changed from one cycle to another by an external signal from a specialized chip. When pipelining is not used (NA HIGH), the address and bus cycle definition data (W/R, D/C, M/IO) remain constant throughout a bus cycle. When the NA pin voltage is low, the address and bus cycle definition signals for the next cycle are in place before the end of the current cycle. This availability is signalled by a LOW output on the address status output pin, which is pin E14 on the 80386DX and pin 16 of the 80386SX.

The BS16 input signal on pin C14 of the 80386DX, active LOW, is used to allow the 80386X to be used either with 32-line or 16-line data buses. When this pin voltage is LOW, the lower half D0–D15 of the data bus is used for all data access, allowing 16-bit bus lines to be used. This is the more usual wiring of the 80386DX chip in many current computers which make use of the AT bus as designed for the 80286 chip. When the BS16 signal is high, the normal 32-bit data lines are used. If the 80387 co-processor is present, the 32-bit bus must be used between the processors, and the BS16 input must not be made active during transfers to or from the 80387. There is no corresponding pin on the 80386SX because the data bus of the 80386SX is only 16 bits wide in any case.

The READY input signal, active LOW, on pin G13 of the 80386DX or pin 7 of the 80386SX indicates to the processor that the current bus cycle is complete and the input data can be latched. The READY signal is ignored on the first bus state in each bus cycle, and is sampled on the remaining states. The system **must** provide a READY signal to ensure that each bus cycle will be correctly terminated.

The HOLD and HLDA actions will be familiar from 8088/8086/80188/80186 and 80286 descriptions. The HOLD pin is D14 on the 80386DX and pin 4 on the 80386SX. The HLDA pin is M14 on the 80386DX and 3 on the 80386SX. A HOLD input, active HIGH, indicates that another chip needs to use the buses, and it must be held high while the other device has bus control. RESET has priority over HOLD at all times. The HLDA pin voltage goes HIGH to indicate acknowledgement of HOLD, signalling that the 80386 has relinquished bus control. When HLDA is high, it is the only signal out of the 80386, all other pins being in the floating state.

During a HOLD, some pins are liable to float to undesired voltages, and the pull–up resistors as indicated in Figure 4.26 are needed to avoid problems. An interrupt on the NMI pin will be remembered for immediate use when a HOLD is relinquished.

Pin	Signal	Pullup value	Purpose
E14	ADS#	20K ± 10%	Pull ADS# negated during HOLD ackn.
C10	LOCK#	20K ± 10%	Pull LOCK# negated in HOLD ackn.

Figure 4.26 *Recommended pull-up resistors for some lines.*

There are three signals that are associated with co-processor use. The PEREQ signal is on pin C8 of the 80386DX and pin 37 of the 80386SX. The BUSY signal is on pin B9 of the 80386DX and pin 37 of the 80386SX. The ERROR signal is on pin A8 of the 80386DX and pin 36 of the 80386SX. PEREQ is active HIGH, BUSY and ERROR are active LOW.

The PEREQ signal is an input from the co-processor which requests data to be transferred to or from the memory by the 80386. When this request is made the co-processor will have completed an instruction which will also be stored in the 80386, so that the 80386 will have the information that is needed to carry out the data transfer in the correct direction (to or from memory) and using the correct memory address. The PEREQ signal is level-sensitive and can be asynchronous.

The BUSY input from the co-processor indicates that the co-processor is still executing an instruction and is unable to accept another. This input is sampled automatically before operations that use the stack (or the WAIT instruction) in order to prevent interfering with a running instruction. The BUSY signal is level sensitive and can be asynchronous.

An alternative use of BUSY performs an internal self-test of the 80386. This occurs if the BUSY input is LOW at the time of the trailing edge of a RESET signal.

The ERROR input from the co-processor occurs if the previous instruction being executed by the co-processor has generated an error that cannot be handled internally (it is not masked by the co-processor's control register). This input is sampled at the start of each co-processor instruction (to find the effect of the previous instruction) and if the voltage is LOW, exception 16 is generated in order to gain access to software routines that will deal with the error. The ERROR signal is level sensitive and can be asynchronous.

There are nine co-processor actions which can execute without generating the exception even if the ERROR level has been pulled low. These are FNINIT, FNCLEX, FSTSW, FSTSWAX, FSTCW, FSTENV, FSAVE, FESTENV and FESAVE.

ERROR has an alternative action. If the error pin voltage is taken LOW at a time of no more than 20

Pin(s)	Signal level during RESET
ADS#	1
D0-D31	Floating
BE0#-BE3#	0
A2-A31	1
W/R#	0
D/C#	1
M/IO#	0
LOCK#	1
HLDA	0

Figure 4.27 *Pin voltages during a 386 RESET.*

CLK2 periods after the trailing edge of RESET, and held LOW for at least one clock cycle following the first bus cycle, the presence of the 80387 will be marked (by setting the ET bit in the CR0 register). If this is not done, then the presence of an 80287 co-processor, or no co-processor, is assumed. The EM and MP bits in the CR0 register must be set by software as usual.

The RESET input, active HIGH, is on pin C9 of the 80386DX and pin 33 of the 80386SX. The pin voltage must be held HIGH for at least 15 CLK2 periods (80 or more if self-test is to be used). When active, any operation in progress will be suspended, all other inputs are ignored, and pins are driven to the idle states as indicated in Figure 4.27. If RESET and HOLD are both activated, RESET takes priority. The RESET signal is level-sensitive and must be synchronized to the CLK2 signal.

The data bus consists of 32 pins on the 80386DX, 16 pins on the 80386SX, active HIGH. The pin assignments were shown in Figure 1.13. The main address bus of the 80386DX consists of lines A2–A31, active HIGH, with the byte enable pins BE0–BE3 used to provide the equivalent of the lower 2 bits of each address, as indicated previously. The 80386SX uses only 24 address pins.

Other outputs

The pins M/IO, D/C and R/W are known collectively as the bus cycle definition outputs, since their signals indicate how the buses are being used. The pins are three-state. These pins are A12, A11 and B12 on the 80386DX and 23, 24 and 25 on the 80386SX respectively. Their functions are best summarized by the table of Figure 4.28 which shows the type of bus cycle indicated by each combination of outputs from these pins. Note that one state does not occur when ADS is not LOW, meaning that it can occur only on

M/IO#	D/C#	W/R#	Bus Cycle Type		Locked?
0	0	0	Interrupt acknowledge		Yes
0	0	1	Forbidden - see Note 1		
0	1	0	I/O Data read		No
0	1	1	I/O Data write		No
1	0	0	Memory instruction read		No
1	0	1	HALT Address=2	SHUTDOWN Address=0	No
1	1	0	Memory data read		Some cycles
1	1	1	Memory data write		

NOTE 1 - The forbidden set can occur in idle states when ADS# is high.

Figure 4.28 *The bus-cycle definition outputs and how they are interpreted.*

an idle bus state, not at any other time. If these outputs are being gated, it may be necessary to use ADS to gate out this particular state.

The LOCK output, active LOW, is on pin C10 of the 80386DX and pin 26 of the 80386SX. This output is issued to prevent any other chip from taking control of the buses, as noted for the other chips in the family. The ADS output, active LOW is on pin E14 of the 80386DX and pin 16 of the 80386SX. This provides address status information, and when LOW indicates that the voltages on the W/R, D/C, M/IO, BE0–BE3 and A2–A31 pins will be valid.

Interrupts and exceptions

The interrupt system for the 80386 processors follows the same pattern as for the other processors in the range. The NMI pin is B8 on the 80386DX and 38 on the 80386SX. The input is active HIGH, edge sensitive, asynchronous, and the voltage on the NMI pin must have been LOW for at least eight CLK2 periods prior to the change, and must be held HIGH for at least eight CLK2 periods. During an NMI interrupt, no further NMIs can be serviced until the processor receives the IRET (Return from Interrupt) command which will be the ending instruction of a service routine. Since the purpose of NMI is to deal with a catastrophic fault, the servicing routine often cannot implement any return so that a second NMI will be ignored. For example, if NMI is used to sense a falling line voltage, the service routine may just have time to save some data, but processing may cease before the end of the routine, depending on how long the machine can operate on the charge on the power supply capacitors.

The INTR interrupt input is on pin B7 of the 80386DX and pin 40 of the 80386SX. This interrupt

INT No.	Function	Related Opcode	Note	Type
0	Divide error	DIV, IDIV	1	Fault
1	Debug exception	any	1	Trap
2	NMI		2	NMI
3	One byte INT	INT	2	Trap
4	Overflow	INTO	2	Trap
5	Array bounds	BOUND	1	Fault
6	Invalid opcode	illegal code	1	Fault
7	Device N/A	ESC, WAIT	1	Fault
8	Double fault	any		Abort
9	Coprocessor O'run	ESC	2	Abort
10	Invalid TSS	JMP, CALL, IRET, INT	1	Fault
11	Segment absent	Segment references	1	Fault
12	Stack fault	Stack references	1	Fault
13	Protection fault	Any memory reference	1	Fault
14	Page fault	Memory access/code fetch	1	Fault
15	Intel reserved			
16	Co-processor error	ESC, WAIT	1	Fault
17-32	Intel reserved			
0-255	Two-byte interrupt	INT n	2	Trap

NOTES: 1 - Return address points to faulty instruction
2 - Return address does not point to faulty instruction
Some Debug exceptions can report a trap on the previous instruction and a fault on the next instruction.

Figure 4.29 *Software interrupts and exceptions list for the 80386.*

can be masked by the IF (interrupt flag) bit in the flags (machine status) register. The response to a HIGH signal on the INTR pin is to perform two interrupt acknowledge bus cycles and following the second acknowledge cycle, to latch an 8-bit address, the interrupt vector, on lines D0–D7 so as to identify the hardware that has caused the interrupt. The input is level-sensitive, asynchronous, and the HIGH input on the INTR pin should remain high until the start of the first interrupt acknowledge bus cycle. Figure 4.29 is the software interrupts and exceptions list for the 80386 chips.

The i486

The i486 features a number of pins which do not correspond to any on the 80386 processors, arising from the considerable enhancements in the i486 design, such as parity checking, internal cache, burst bus use and better bus arbitration. In addition, the address pins of the i4586 are input/output capable, unlike the address pins of the 80386 chips. This latter facility is required for cache invalidation checking, noted later in this section.

The DP0–DP3 input/output pins are used for data parity, using one pin for each byte on the data bus. Even parity signals are generated on these pins at each write, and these parity signals should be returned on a read action so that the processor will indicate correct parity check status. This parity check does *not* cause an interrupt, and the operating system needs to read the parity status output and take action if an error is found. The parity status output is on the PCHK# pin, which goes LOW to indicate a parity error on a read of code, data or I/O. The PCHK# pin is never floated. If parity checking is not implemented, the DP0–DP3 pins should be pulled high.

The i486 uses the normal LOCK# signal to indicate that the bus must not be released between a read and a write cycle. There is also a pseudo-lock, or PLOCK# pin which is used to maintain control of the buses throughout reads and writes of operands that use more than 32 bits. PLOCK# will be driven LOW during reads and writes of long floating-point numbers (64 bits) and cache line fills of 128 bits. Asserting PLOCK# prevents a HOLD from being acknowledged until the PLOCK is released, and neither RDY# nor BRDY# will release the buses until the transfer of bits has been completed. See also BLAST# and KEN#.

The ADS# output is a bus control signal that indicates that the address and bus cycle definition signals are valid. ADS# will normally be asserted in the first clock period of a bus cycle and will be inactive in the subsequent clock periods or when the bus is idle or on hold. The signal can be used by external devices as an indication that a bus cycle has started and that the bus cycle definition pins can be sampled on the next rising edge of the clock pulse.

The RDY# signal is asserted to indicate that the current bus cycle is complete. On a read cycle, the appearance of RDY# indicates that valid data has been read; on a write cycle it indicates that the data has been written. The signal is ignored when the bus is idle, and also at the end of the first clock cycle. There is no internal pull-up resistor.

The BOFF# (backoff) input forces the i486 to release control of the buses on the next clock period, floating the pins as if a HOLD were being executed. Unlike a HOLD, the response to BOFF# is immediate; the current bus cycle is not completed, and HLDA is not asserted. BOFF# has higher priority than RDY# or BRDY#. Bus hold continues until BOFF# is negated, at which time the aborted bus cycle can be completed. In this respect, the action of

BOFF# is the same as that of inserting wait states. Any data arriving at the CPU during active BOFF# will be ignored.

The BS16# and BS8 inputs allow the chip to be used with buses smaller than 32-bit, and these pin inputs are sampled at each clock cycle. Asserting either of these pins will cause a data read to take place in two or four stages respectively. Both pins have internal pull-up resistors. The A20M# pin is the address bit 20 mask, and asserting this input causes address bit 20 to be masked in any memory cycle or cache look-up. This causes the i486 to emulate the 1M address space of the 8086.

Cache specific signals

The AHOLD and EADS# inputs are used in cache invalidation cycles. Cache invalidation cycles are required to ensure that the contents of the internal cache are consistent with the contents of the external memory, so that altering part of the external memory which happens to be cached will not result in any discrepancy in data. When a memory write is made (by an external device) to a part of main memory whose contents are also in the cache, the cache contents are automatically invalidated – if the external memory is altered by the i486, of course, this will involve changing the cache in the process so that invalidation can never occur in this way.

When an external write to main memory occurs, the AHOLD (address hold request) is asserted to force the i486 to release the address bus. The external system will then assert EADS# to indicate that a valid address is on the bus. The i486 will read this address and compare it with cached address ranges. If the address corresponds to a cached piece of memory, the cache entry is invalidated – note that this action requires the use of address pins for **input**. Once a part of the cache has been invalidated it will not be used, and will be replaced next time the cache is used.

Cache control is carried out using KEN#, FLUSH#, PWT and PCB. The KEN# signal is used to determine whether the data for the current cycle is cacheable or not. If KEN# is asserted and a memory read cycle is being performed, the cycle will be used as a cache fill, and a cache line of 16 bytes will be read (byte-enable pins will be ignored). KEN# is provided with an internal pull-up resistor. The FLUSH# input forces all of the cache contents to be abandoned if this input is asserted for one clock period. The signal is asynchronous, and correct set-up and hold times must be observed.

The PWT and PCD output signals are derived from the two user attribute bits in the page table entry (see later, this section). When paging is enabled, these pin signals correspond to bits 3 and 4 of the page table entry. When paging is disabled, the bits are 3 and 4 in the CR3 register.

Floating-point system signals

The FERR# output is asserted when an unmasked floating-point error happens, and the signal corresponds to the ERROR# signal of the 80387, see Section 6. The signal can be used by external circuitry for reporting such errors. The IGNNE# pin input can be asserted to force the floating point unit to ignore a numerical error and continue operations.

Burst bus use

In a 'normal' bus cycle, as in the 80386, a new data item can be read on each pair of clock cycles. Burst use of buses allows a set of data items to be read, after a first item read, on each clock period rather than on alternative clock periods, doubling the speed of input. During a write, only 32 bits can be burst, so that burst writing is effective only when BS16# or BS8# are active.

In a burst action, an address is provided and ADS# is asserted as for a normal bus action, but the burst last (BLAST#) signal is held negated in the second clock period. If the external system is capable of supporting a burst cycle, it will acknowledge by asserting the burst ready (BRDY#) input. To make this possible, each burst must start at a location XXXXXXX0H and end at XXXXXXXFH, making it easier for external devices to calculate the addresses for data transfer.

Cache organization

Figure 4.30 illustrates the physical organization of the cache as a set of four 2K blocks, each block having a 21-bit set of tags and with a control block. Each data block consists of 128 lines, each of 16 bytes, and is addressed in the TR5 register using 7

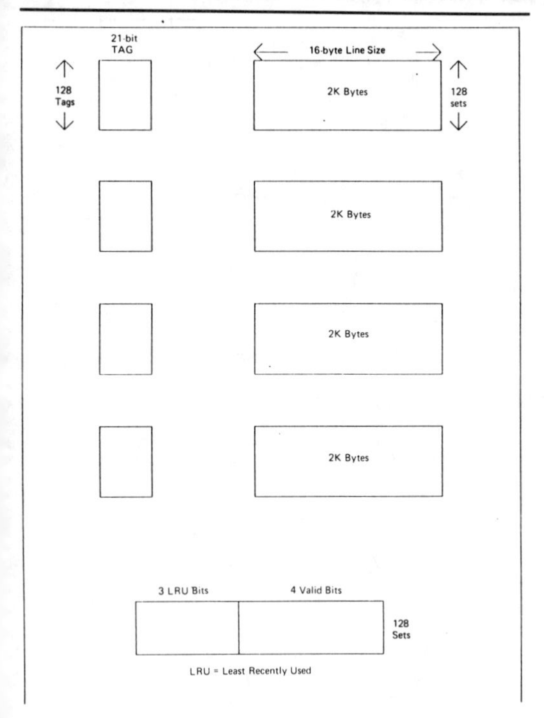

Figure 4.30 *The physical layout diagram for the cache memory.*

bits for the set number and 2 bits for the block number. For each 16-byte line in the data blocks there will be a 21-bit tag number consisting of a 17-bit linear address and 4 protection bits, as in the TLB tags (Section 3). The control block contains 128 sets of 7 bits, of which 3 bits are used to implement the LRU (least-recently used) system for replacement, and 4 bits used to indicate validity.

One set in a cache consists of 16 bytes, or 4 *lines* of 32-bit data, and the validity of a set is tested by testing each line, using the strategy indicated in Figure 4.31. If any line is non–valid the complete set is made available for replacement, otherwise lines can be replaced on the basis of not being recently used, making use of the LRU bits which are updated each time a line is replaced or used. When the cache is

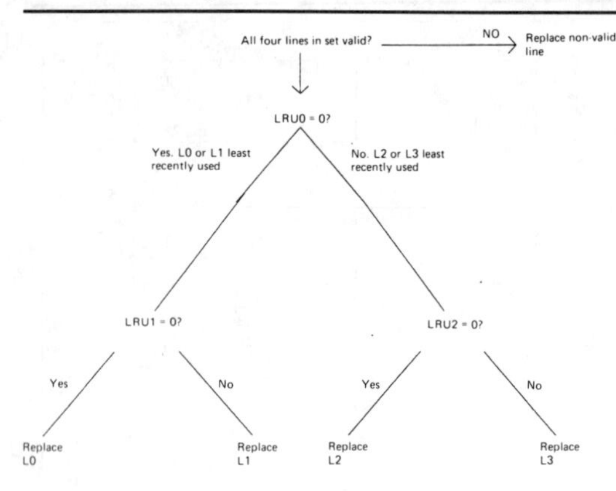

Figure 4.31 *A diagram showing the cache strategy.*

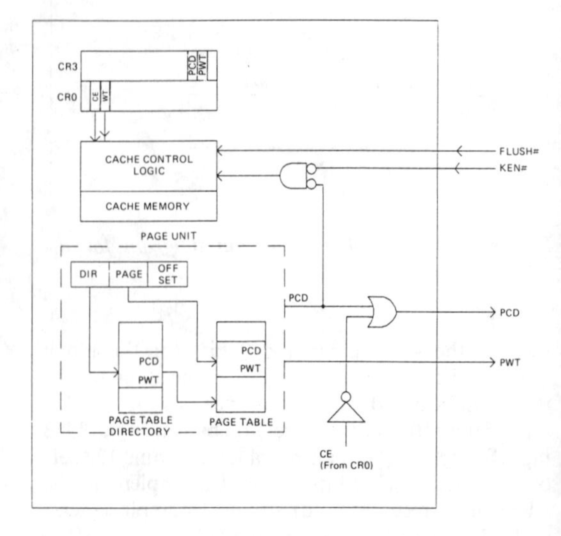

Figure 4.32 *Block diagram for the cache control unit.*

flushed (or the CPU reset) all of the LRU bits are cleared.

The block diagram for the cache control is shown in Figure 4.32. The CR0 and CR3 registers are closely involved, using the PCD and PWT bits in

CE	WT	Operating mode
0	0	Cache fills, write-through, invalidates all disabled
0	1	Fills disabled, write-through and invalidates enabled
1	0	INVALID CODING - will cause protection fault, code 0
1	1	Fills, write-through and invalidates all enabled

Figure 4.33 *Interpreting the CE and WT bits in the CR0 register.*

CR3. The PCD bit determines the cacheability of each page, using PCD=0 to enable caching. The PWT bit determines the writing method for external caches – for PWT=1 the current page uses write–through, for PWT=0 the system is write-back. This bit is ignored by the i486 because the built-in cache is write through, meaning that an address in the cache will be written in addition to an external address when data is written.

The control of the cache is by way of the CE and WT bits in register CR0 (Figure 4.33). When both of these bits are cleared, the cache can be completely disabled by carrying out a flush action (note that cache reads can still be performed with both bits cleared). The CE=1, WT=1 mode is the normal operating mode enabling cache fills, write-through and invalidation to be carried out.

5 Timing of signals

Timing of microprocessor signals is conventionally shown in a diagram which incorporates a large number of different timing situations in one set, and which can be notoriously difficult to unravel. In this section, timing patterns will be described both with simplified diagrams and in words, concentrating on the **normal** bus timings first and the timings of such events as waits and halts later. The timing diagrams provided by the manufacturer will have to be referred to for applications that are non–standard or which involve the use of non-Intel support chips, but the treatment provided here should be adequate for 99 per cent of users. **Remember** that timing diagrams indicate only the **sequence** of events and are **never** to scale, so that all times should be obtained from the Intel specifications.

As before, the timings of the different processors will be dealt with separately, but referring back to the earlier units in situations where timings are almost identical.

For all timing signals, the following assumptions are made:

- Test points for time measurement are at the 1.5 V level.
- All input signals must switch between 0.45 V and 2.4 V.
- Timings shown assume a 'standard' value of capacitive load, typically 30–100 pF.

The 8088/8086 timings

The source of clock signals is assumed to be the 82C84A chip, whose output signal is asymmetrical with a 33.33 per cent duty cycle. The typical waveform is shown in Figure 5.1. For more details of the 82C84A chip, see Section 7. Note that timing diagrams for the processors, by convention, show the

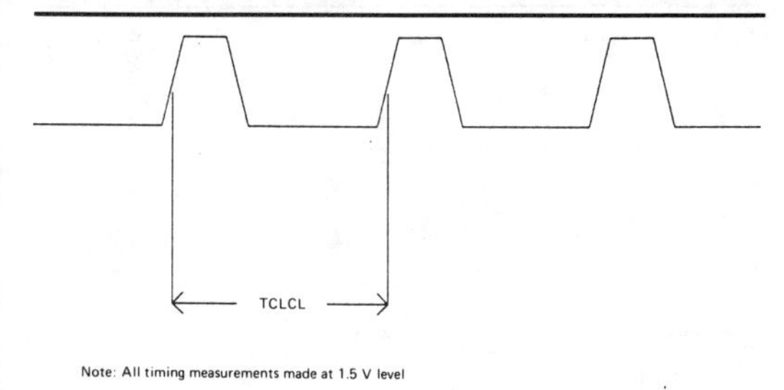

Figure 5.1 *Typical clock signal waveform for 8088/8086 processors.*

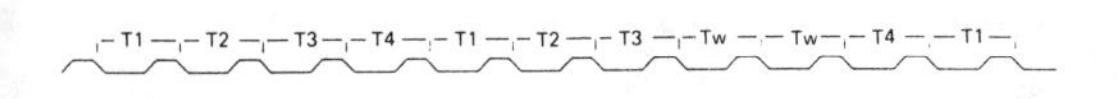

Figure 5.2 *Wait cycles inserted between T3 and T4 cycles.*

CLK wave-form as symmetrical. This should not cause confusion if it is clear from which edge of the clock pulse timing is taken.

The CLK signal is internally divided in the 8088/8086 to half of the input frequency, but timings are indicated in terms of the input CLK signals because both phases of the internal clock are used. Each processor bus cycle consists of at least **four** CLK cycles, labelled as T1, T2, T3 and T4. If a wait is inserted, a number of CLK cycles of waiting will be inserted between T3 and T4 (Figure 5.2). In a normal bus cycle of four CLK periods:

- The address is put out during T1.
- T2 is used for changing the direction of the bus for reading.
- Data transfer occurs during T3 and T4.
- The ALE signal is present during T1 to enable address latching.

Figure 5.3 shows this conventional bus cycle for a minimum system in diagrammatic form. On the 8088, three sets of pins are used, of which the A8–A15 set consist only of address data. These will become valid some time after the rise of the ALE signal, as will the multiplexed address lines, all of which provide address data into the T2 period. The

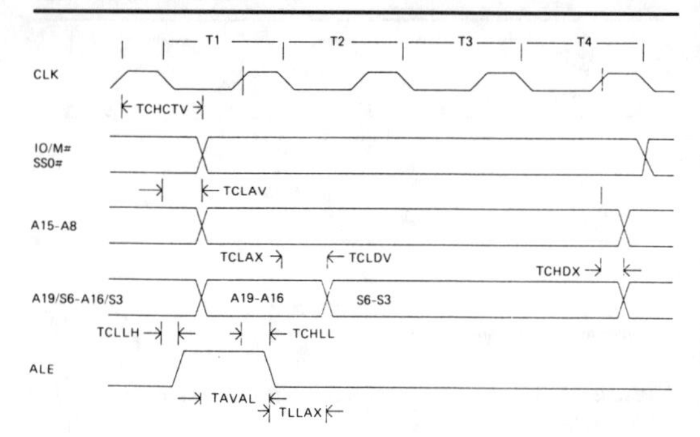

Figure 5.3 *A conventional bus cycle for a minimum system.*

important point here is that valid address data will exist on all of these lines by the time when the ALE trailing edge occurs and the data on the A0–A7 and A16–A19 lines can be latched. The data on the A8–A15 lines remains valid until the leading edge of the next T1 period.

During the T3 and T4 cycles (and part of T2), the status information signals S3–S6 are available on the lines that are shared with A16–A19, and during the T4 period, the data signals D0–D7 exist on the lines that are shared with A0–A7. The time between a valid address being established and the appearance of data on the data lines during a read action will depend on the memory access time rather than on the processor timing. If slow memory is being used, one or more wait periods may have to be inserted between T3 and T4 in each bus cycle that involves a memory read or write.

Wait states are controlled by the READY signal. This can be asserted LOW at any time in T2, even near the end of T1, and if LOW in T3 will cause a wait period to be added before T4. Normal bus action will resume at about one clock period following the READY signal going HIGH. This action is summarized in Figure 5.4.

On a bus cycle for a read, the read strobe (RD) will go low towards the end of T2, **but not until the bus has floated following the end of address data**. The read strobe will remain low in T3 and also during any wait periods that are required for reading slow memory,

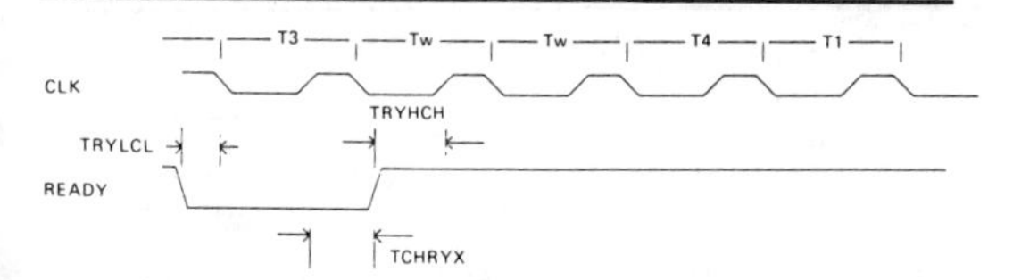

Figure 5.4 *Using READY to impose wait states in a bus cycle. Normally, READY would be LOW* before *the start of T33 and would be put HIGH to end a cycle or held LOW to impose wait states.*

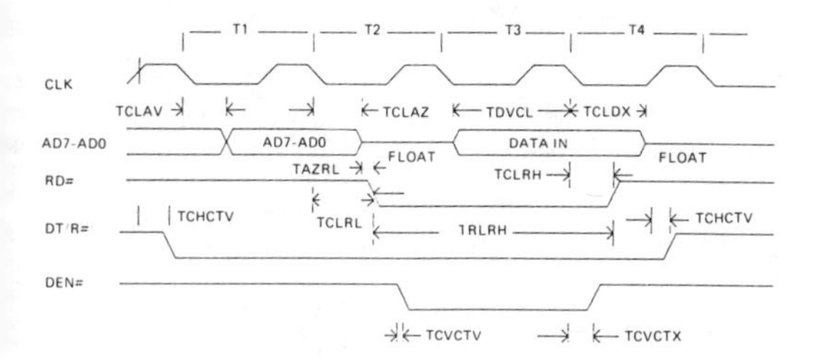

Figure 5.5 *The bus cycle for a read action, showing only the changes compared with Figure 5.3.*

ending in the T4 part of the bus cycle. Valid data will exist on the D0–D7 lines by the end of the RD strobe pulse. This part of the action is summarized in Figure 5.5. The data bus is reserved for use once the read strobe has gone low so as to allow for the use of faster memory. It is unusual to require wait states for slow memory on 8088 machines, since most memory is nowadays fast enough to cope with the speed of the 8088 processor.

On a bus cycle for a write, the WR strobe signal will go LOW towards the end of a T2 period and will be held low until the end of T3 or the end of any wait period that follows T3, whichever is later. When WR is low, the address lines A0–A7 will change to become data output lines D0–D7, about half of a clock period later than the falling edge of ALE. The data remains valid for the latter part of T2, the whole of any wait periods and until approximately halfway through T4. These timings are indicated in Figure 5.6.

For a minimum system, the DEN (data enable) pin

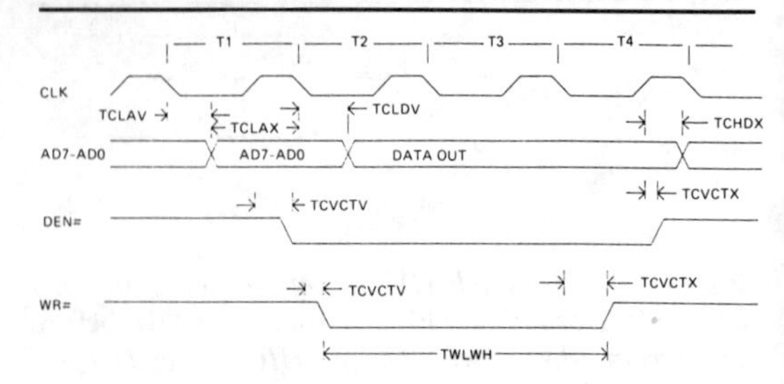

Figure 5.6 *The bus cycle for a write action, showing only the changes compared with Figure 5.3.*

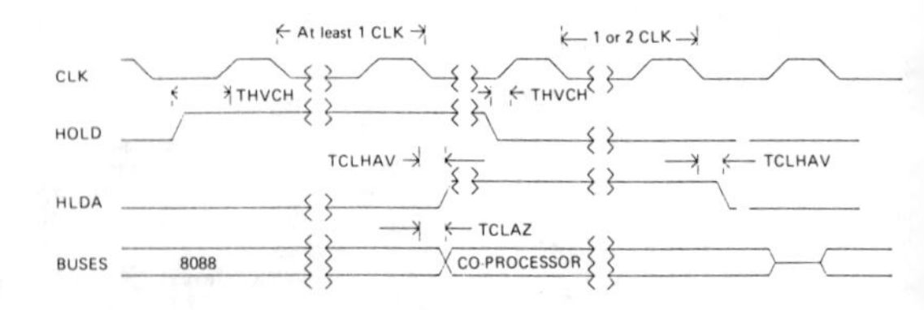

Figure 5.7 *The timing of HOLD and HLDA for a MAXIMUM* system.

voltage can be used to enable a data transceiver, if this type of chip is used (a very small-scale system might not use a data transceiver). The DEN voltage will go active LOW for a read or write cycle, but with different timings. For a data read, DEN goes low during the high-voltage part of T2 and rises again in the high-voltage portion of T3 or Tw. For a data write, DEN goes low earlier, prior to the low-to-high transition of T2, and rises again later at the start of the T4 period. These timings are shown in Figure 5.5 and 5.6.

Figure 5.7 shows the timing of the HOLD/HLDA signals. The HOLD signal is shown as reaching its active HIGH voltage before the trailing edge of a clock pulse, and the HLDA output goes high some time later, following the falling edge of the T4 pulse of the bus cycle. By this time the address/data buses will have been floated along with DEN, RD, WR, S7, DT/R, S0–S2 and LOCK. When the HOLD voltage drops again, the HLDA will drop a few clock cycles later, and the buses will be released – full release of the buses is ensured by the start of the next bus cycle with one T4 cycle following the release of the HLDA signal.

Maximum mode timing changes

The timings that have been illustrated for minimum mode on the 8088 apply as far as the use of the data and address buses is concerned to the maximum mode, but the nature of several control signals is altered. It is assumed that in maximum mode, the 8088 will be used along with the 8288 bus controller chip, and this chip will provide the DT/R, MRDC or IORC, DEN, AMWC or AIOWC, INTA and other signals that a maximum system is likely to need, some of which are provided directly from 8088 pins in a minimum system. In a maximum system, these pins are used for status, request grant, queue status and lock signals, and it is the timing of these that distinguishes the maximum mode. The lock timing will be discussed later when considering interrupt timings.

The status pins S0–S2 provide valid information from just after the start of T1 until just the start of T3 or whatever wait state precedes the T4 period. Unless a software HALT is being implemented, at least one of these pins will be at LOW voltage during this time, so that diagrams illustrate the logical AND of these three pin signals (Figure 5.8).

The request/grant sequence (Figure 5.9), can start in any part of the bus cycle with a pulse issued from the co-processor to the RQ pin, timed from midway though one clock period to midway through the next. This requests the use of the buses, and when the 8088 has floated the buses, it will issue another pulse from

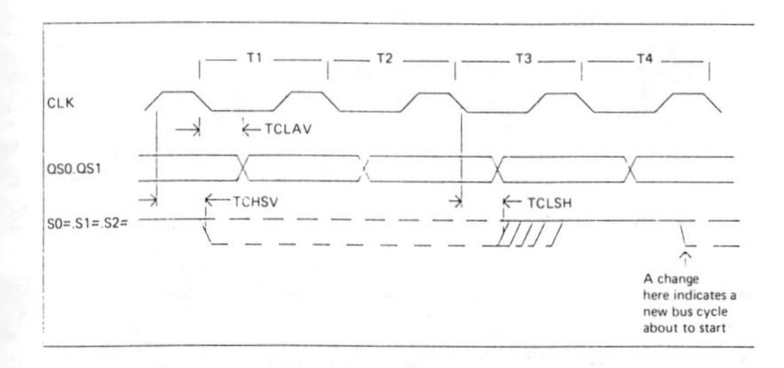

Figure 5.8 *The timing of the signals from the S0–S2 pins.*

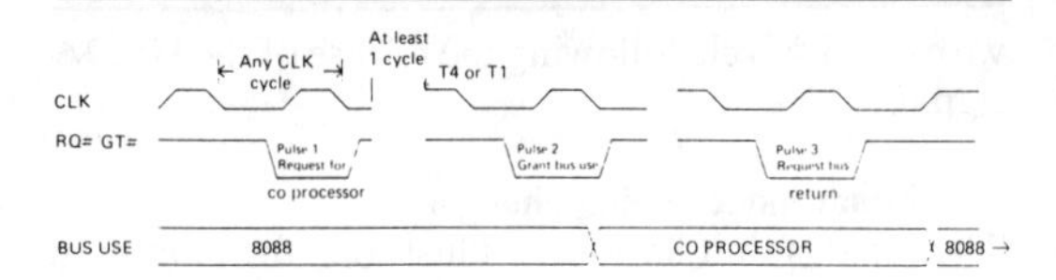

Figure 5.9 *The request/grant sequence.*

the same pin and with the same timing to the co-processor to acknowledge that the buses are now free for use. From the start of this process, the clock pulses are not acting in the normal T1–T4 pattern. The buses are then used by the co-processor, and when this use has been completed, a third pulse will be issued from the co-processor to the RQ pin to indicate the end of the bus use. The following clock period is then treated as a T4 cycle, and a normal bus cycle will start on the period following the T4.

The queue–status output signals are available during each period, and should be sampled during the high voltage part of a clock period. The information refers to the state of the queue for the following period.

Interrupt sequences

The interrupt sequences (Figure 5.10), follow the same pattern in both minimum and maximum modes with the exception of the LOCK output signal, which is used only in maximum mode. The interrupt acknowledge sequence uses two bus cycles and two consecutive INTA pulses. In a minimum system, there will be two ALE pulses, with the INTA voltage going low at the end of each ALE pulse. On a

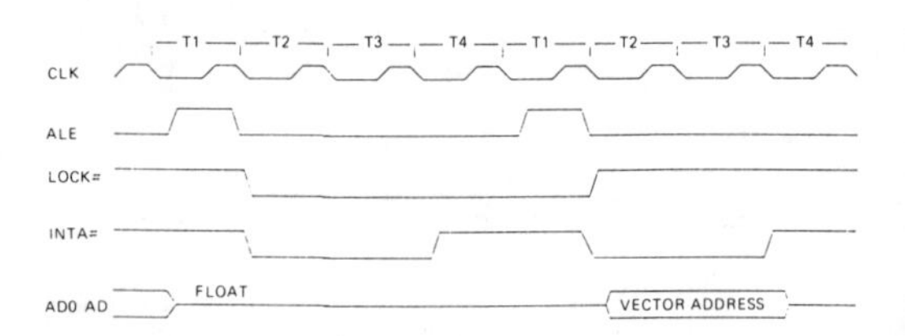

Figure 5.10 *The interrupt sequence for minimum mode. Maximum mode sequences depend on the outputs from the 8288 chip.*

maximum system, the LOCK output will go low at the trailing edge of T1, coincident with the first INTA pulse, and LOCK will go high again at the end of the second T1 cycle, initiating the second INTA pulse. The AD0–AD7 bus will be floated from midway through the first T1 period until just after the falling edge of the second INTA pulse. At this point, the bus will be used to provide the 8-bit vector address for the interrupt source, terminating at the trailing rising edge of INTA. Following the T4 pulse, normal bus actions will resume so that the service routine for the interrupt can be run.

8086 differences

The most notable difference between 8086 and 8088 timing is that the main multiplexed bus is now AD0–AD15, whose timing is as for the AD0–AD7 part of the 8088 bus. The A16–A19 portion of the bus is identical in both processors. In an interrupt cycle, only the lines AD0–AD7 are used for the interrupt vector. A minor difference occurs in the ALE timing for minimum mode when a software HALT is being implemented. On the 8088, the ALE is delayed by one period so as to allow the status bits to be latched by the ALE signal; this delay is not implemented on the 8086.

Figure 5.11 is a summary of the timing requirements, excluding clock signals, for the various 8088 chips and Figure 5.12 is a corresponding summary for the 8086. These summaries use the Intel abbreviations for the various critical times, and also brief explanations of the times. Remember the convention that times are measured between the 1.5 V points in rising or falling wave-forms.

The 80188/80186 processors

The source of clock signals for these chips is an internal oscillator, and timing is illustrated in terms of the CLKOUT signal on pin 56. The same sequence of four clock periods T1–T4, with wait states inserted between T3 and T4, is used for the 80188/80186 processors as for the earlier 8088/8086. The

NOTE: All times in nanoseconds, ns.

I - Essential timing Requirements.

Symbol	Parameter	8088 Min	8088 Max	8088-2 Min	8088-2 Max	Notes
TDVCL	Setup time, data in	30		20		
TCLDX	Hold time, data in	10		10		
TR1VCL	RDY setup for 8284	35		35		1,2
TCLR1X	RDY Hold time for 8284	0		0		
TRYHCH	Setup time, READY, 8088	118		68		
TCHRYX	Hold time, READY, 8088	30		20		
TRYLCL	READY inactive to CLK	-8		-8		
TINVCH	Setup time for recognition	30		15		2
TGVCH	Setup time, RQ/GT	30		15		
TCHGX	Hold time, RQ/GT	40		30		
TILIH	Input rise time (not CLK)		20		20	3
TIHIL	Input fall time (not CLK)		12		12	4

Notes: 1. Refers to use of 8284/8288
2. To guarantee recognition at next CLK
3. From 0.8 V to 2.0 V levels
4. From 2.0 V to 0.8 V levels

II - Required Timing Responses

Symbol	Parameter	8088 Min	8088 Max	8088-2 Min	8088-2 Max	Notes
TCLML	Command active delay	10	35	10	35	1
TCLMH	Command inactive delay	10	35	10	35	1
TRYHSH	READY active to passive		110		65	2
TCHSV	Status active delay	10	110	10	60	
TCLSH	Status inactive delay	10	130	10	70	
TCLAV	Address valid delay	10	110	10	60	
TCLAX	Address Hold time	10		10		
TCLAZ	Address float delay	*	80	*	50	3
TSVLH	Status valid to ALE High		15		15	1
TSVMCH	Status valid to MCH High		15		15	1
TCLLH	CLK Low to ALE valid		15		15	
TCLMCH	CLK Low to MCE		15		15	1
TCHLL	ALE inactive delay		15		15	1
TCLMCL	MCE inactive delay		15		15	1
TCLDV	Data valid delay	10	110	10	60	
TCHDX	Data Hold time	10		10		
TCVNV	Control active delay	5	45	5	45	1
TCVNX	Control inactive delay	10	45	10	45	1
TAZRL	Address float to read active	0		0		
TCLRL	RD# active delay	10	165	10	100	
TCLRH	RD# inactive delay	10	150	10	80	
TRHAV	RD# inactive-next addr. act.	*		*		4
TCHDTL	Direction control active dely		50		50	
TCHDTH	Dir. ctrl. inactive delay		30		30	
TCLGL	GT active delay		85		50	
TCLGH	GT inactive delay		85		50	
TRLRH	RD# width	*		*		5
TOLOH	Output rise time		20		20	
TOHOL	Output fall time		12		12	

NOTES: 1. 8288/8284 signals
2. Applies only to T3 and Tw states
3. Minimum time is equal to TCLAX value
4. Minimum is TCLCL-45 for 8088, TCLCL-40 for 8088-2
5. Minimum is 2TCLCL-75 for 8088, 2TCLCL-50 for 8088-2
* Depend on other timings

Figure 5.11 *A summary of timing requirements for 8088, maximum mode. The heading of 8088-2 covers the later CHMOS versions. Symbols are as shown in timing diagrams. Note that some timings depend on other chips, a few timings depend on timings of other actions.*

NOTE: All times in nanoseconds, ns.

I - Essential timing Requirements.

Symbol	Parameter	8086		8086-1		8086-2		Notes
		Min	Max	Min	Max	Min	Max	
TDVCL	Setup time, data in	30		5		20		
TCLDX	Hold time, data in	10		10		10		
TR1VCL	RDY setup for 8284	35		35		35		1
TCLR1X	RDY Hold time for 8284	0		0		0		1
TRYHCH	Setup time, READY, 8088	118		53		68		
TCHRYX	Hold time, READY, 8088	30		20		20		
TRYLCL	READY inactive to CLK	-8		-10		-8		
TINVCH	Setup time for recognition	30		15		15		2
TGVCH	Setup time, RQ/GT	30		15		15		
TCHGX	Hold time, RQ/GT	40		20		30		
TILIH	Input rise time (not CLK)		20		20		20	3
TIHIL	Input fall time (not CLK)		12		12		12	4

Notes: 1. Refers to use of 8284/8288
2. To guarantee recognition at next CLK, asynchronous signal only
3. From 0.8 V to 2.0 V levels
4. From 2.0 V to 0.8 V levels

II - Required Timing Responses

Symbol	Parameter	8086		8086-1		8086-2		Notes
		Min	Max	Min	Max	Min	Max	
TCLML	Command active delay	10	35	10	35	10	35	1
TCLMH	Command inactive delay	10	35	10	35	10	35	1
TRYHSH	READY active to passive		110		45		65	2
TCHSV	Status active delay	10	110	10	45	10	60	
TCLSH	Status inactive delay	10	130	10	55	10	70	
TCLAV	Address valid delay	10	110	10	50	10	60	
TCLAX	Address Hold time	10		10		10		
TCLAZ	Address float delay	*	80	10	40	*	50	3
TSVLH	Status valid to ALE High		15		15		15	1
TSVMCH	Status valid to MCH High		15		15		15	1
TCLLH	CLK Low to ALE valid		15		15		15	1
TCLMCH	CLK Low to MCE		15		15		15	1
TCHLL	ALE inactive delay		15		15		15	1
TCLMCL	MCE inactive delay		15		15		15	1
TCLDV	Data valid delay	10	110	10	50	10	60	
TCHDX	Data Hold time	10		10		10		
TCVNV	Control active delay	5	45	5	45	5	45	1
TCVNX	Control inactive delay	10	45	10	45	10	45	1
TAZRL	Address float to read active	0		0		0		
TCLRL	RD# active delay	10	165	10	70		100	
TCLRH	RD# inactive delay	10	150	10	60		80	
TRHAV	RD# inactive-next addr. act.	*		*		*		4
TCHDTL	Direction control active dely		50		50		50	1
TCHDTH	Dir. ctrl. inactive delay		30		30		30	1
TCLGL	GT active delay	0	85	0	38	0	50	
TCLGH	GT inactive delay	0	85	0	45	0	50	
TRLRH	RD# width	*		*		*		5
TOLOH	Output rise time		20		20		20	
TOHOL	Output fall time		12		12		12	

NOTES:
1. 8288/8284 signals
2. Applies only to T3 and Tw states
3. Minimum time is equal to TCLAX value
4. Minimum is TCLCL-45 for 8086, TCLCL-35 for 8086-1, TCLCL-40 for 8086-2
5. Minimum is 2TCLCL-75 for 8086, 2TCLCL-40 for 8086-1, 2TCLCL-50 for 8086-2
* Depends on other timings

Figure 5.12 *A summary of timing requirements for 8086, maximum mode. Symbols are as shown in timing diagrams.*

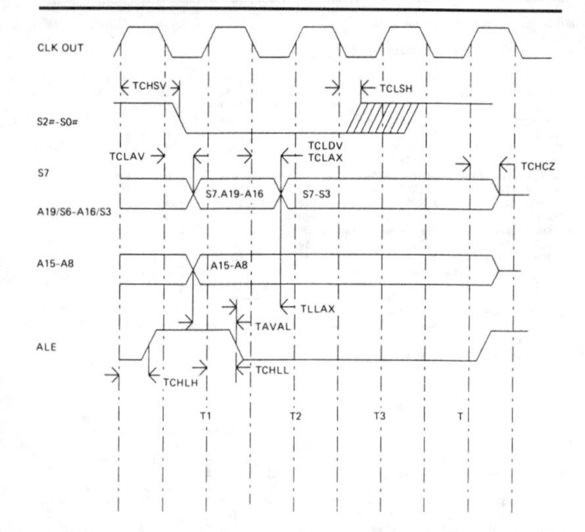

Figure 5.13 *Major bus cycle timings for 80188/80186, 80186 buses illustrated.*

CLKOUT wave-form is symmetrical, with a 50 per cent duty ratio.

Figure 5.13 shows the bus cycle of timings that are common to all operations. There is no minimum/maximum system switch for these chips, since they contain all the elements of a maximum-mode 8088/8086 system. The A8–A15 bus is included only for the 80188, since on the 80186 these functions are on the AD8–AD15 lines which are dealt with separately in the various read or write cycles. Note that the ALE signal is active earlier than in the 8088/8086 processors, so that ALE is established before the start of a T1 period and ends a short time (typically 35 ns) after the low-to-high transition in the T1 period.

The status signals S0–S3 are valid during the T1 and T2 cycles, but are inactive during the T3 cycle. The S7 signal is HIGH, and the A16–A19/S3–S6 signals are LOW prior to the transition in the first part of the T1 period. During the rest of T1, the address bits A16–A19 are valid on these multiplexed address lines, and the S7 signal is valid on its pin. The S7 signal continues to be valid through the rest of T2 and T3 to the end of T4, and in this latter part of the cycle, the multiplexed lines carry the S3–S6 signals. On the 80188 only, the A8–A15 lines hold a valid address from a point in the low-voltage part of T1 to the end of T4. The timing of the multiplexed AD8–AD15 lines of the 80186 is different, see later.

Wait states are controlled by the ARDY or SRDY signals. The ARDY input is the asynchronous ready, which will insert a wait state when put LOW. The rising edge can be asynchronous, but the falling edge must be synchronized to the processor clock. Tying the pin to V_{cc} ensures that wait states are never used. If the pin is tied low, then control passes to SRDY. The SRDY synchronous ready input has to be synchronized externally to the clock, and will insert a wait state when taken LOW. This timing is summarized in Figure 5.14.

On a bus cycle for a read, the read strobe (RD) will go low towards the end of T2, **but not until the bus has floated following the end of address data**. The read strobe will remain low in T3 and also during any wait periods that are required for reading slow memory, ending in the T4 part of the bus cycle. Valid data will exist on the D0–D7 lines of the 80188 and the D0–D15 lines of the 80186 by the end of the RD strobe pulse. This part of the action is summarized in Figure 5.15. The DT/R line will go LOW at the start of the T1 cycle, and HIGH again in the T4 cycle to enable data to be received through a data transceiver. The DEN line goes LOW in the T2 cycle and HIGH again in the T4 cycle to enable the data transceiver chip. In addition, the memory and peri-

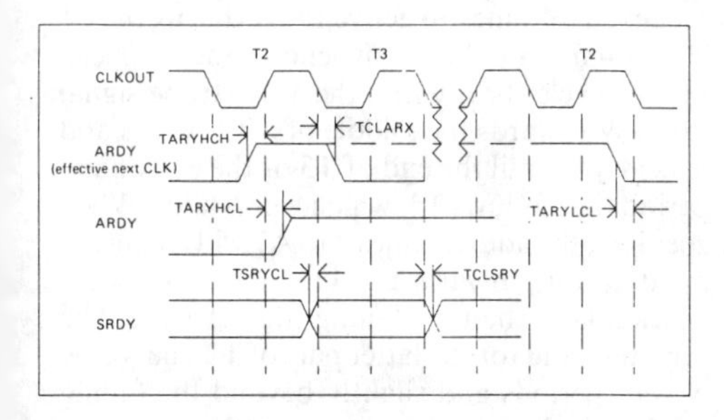

Figure 5.14 *The synchronous READY signal SRDY and the asynchronous READY, ARDY. Either of these can be used to terminate a cycle.*

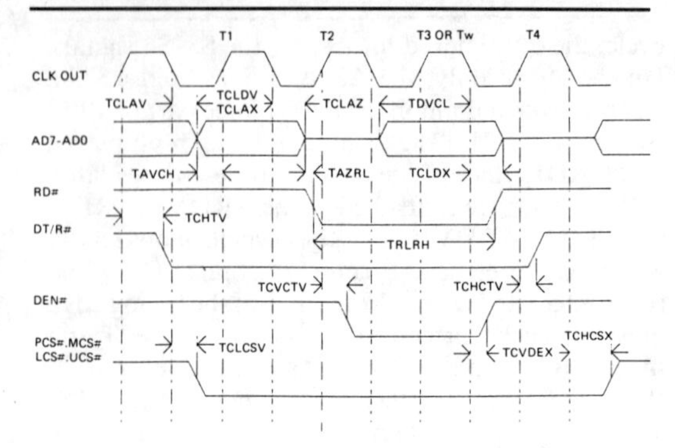

Figure 5.15 *A read bus cycle for the 80188/80186, 80186 buses illustrated.*

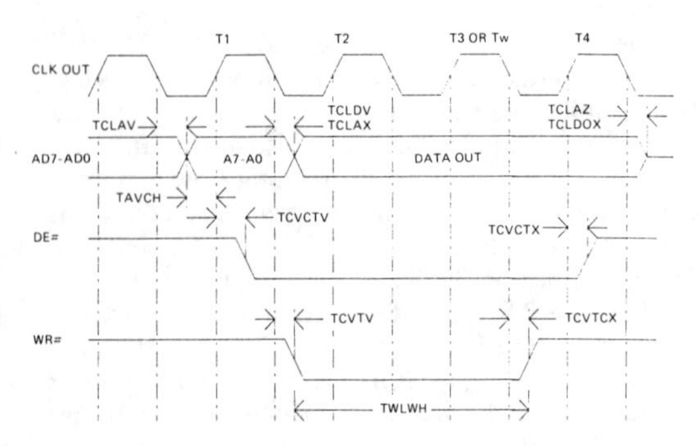

Figure 5.16 *A write bus cycle for the 80188/80186, 80186 buses illustrated.*

pheral chip enable lines go active LOW during the T1 period and remain LOW to the end of the T4 cycle.

On a bus cycle for a write, the WR strobe signal will go LOW towards the middle of a T2 period and will be held low until the end of T3 or the end of any wait period that follows T3, whichever is later. When WR goes low, the address lines A0–A7 will change to become data output lines D0–D7, about half of a clock period later than the falling edge of ALE. The data remains valid for the latter part of T2, the whole of any wait periods and slightly beyond the falling edge of T4. The DEN voltage goes LOW just before the end of the T1 cycle and remains LOW until midway through the T3 or the last Tw period. These timings are indicated in Figure 5.16.

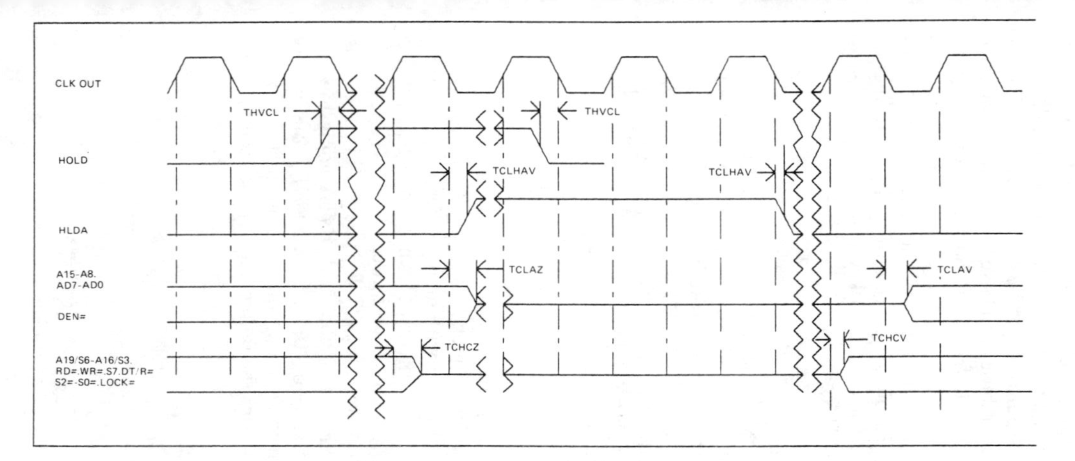

Figure 5.17 *Timing of HOLD and HLDA signals for the 80188/80186.*

Figure 5.17 shows the timing of the HOLD/HLDA signals. The HOLD signal is shown as reaching its active HIGH voltage before the trailing edge of a clock pulse, and the HLDA output goes high some time later, following the falling edge of the T4 pulse of the bus cycle. By this time the address/data buses will have been floated along with DEN, RD, WR, S7, DT/R, S0–S2 and LOCK. When the HOLD voltage drops again, the HLDA will drop a few clock cycles later, and the buses will be released – full release of the buses is ensured by the start of the next bus cycle with one T4 cycle following the release of the HLDA signal.

The timers

The timers of the 80188 operate in a way that is illustrated in Figure 5.18. In this diagram, the input clock signal is shown also, because it is used for the timing action. The TIMERIN signals must go high prior to the rising edge of one of the CLKIN pulses, and the TIMEROUT signal will follow after 2.5–6.5 clock periods later, depending on the form of timing that is being used. The TIMEROUT signal will then remain high until the timing period is ended.

Interrupt sequences

The interrupt sequences (Figure 5.19), follow the same pattern, but with timing differences, as for the 8088/8086. The interrupt acknowledge sequence uses two bus cycles and two consecutive INTA pulses. In the second cycle, the DT/R voltage will go low just prior to the T1 period, with INTA going low just after the start of T1. The AD0–AD7 bus will be floated from midway through the first T1 period until just after the falling edge of the second INTA

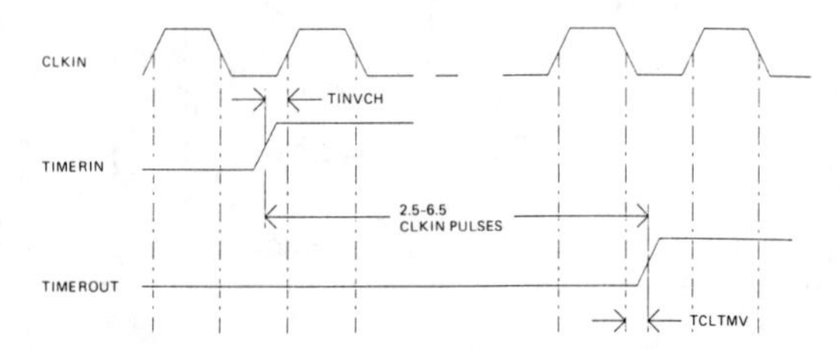

Figure 5.18 *The timer IN and OUT actions relative to CLKIN.*

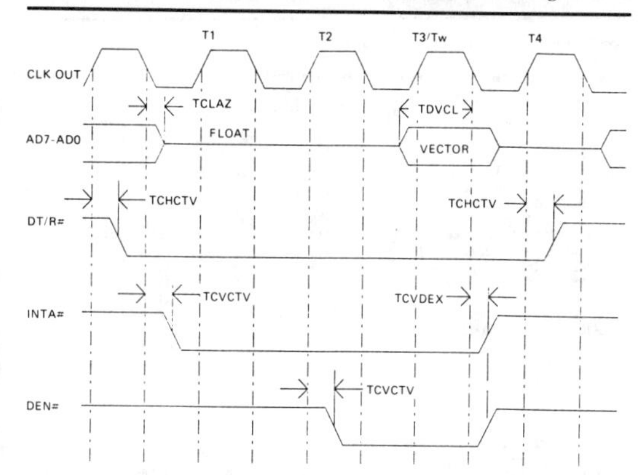

Figure 5.19 *Interrupt sequences for the 80188/80186.*

pulse. At this point, the bus will be used to provide the 8-bit vector address for the interrupt source, terminating at the trailing rising edge of INTA. The DEN voltage goes LOW midway through T2 and rises again along with INTA in T4. Following the T4 pulse, normal bus actions will resume so that the service routine for the interrupt can be run. Figure 5.20 summarizes the timing requirements of the 80186/80188 chips.

The 80286 timings

The timing pattern of the 80286 follows very closely the patterns established by its predecessors, but there are some changes in nomenclature that can be confusing. As before, the clock input frequency is divided by two, and the resulting frequency is referred to as the **processor clock**, with the undivided frequency termed the **system clock**. Each processor clock cycle is divided into two **phases**, one for each complete system clock cycle (Figure 5.21). A bus cycle consists of two processor clock cycles, corresponding to the T1–T4 cycle of the earlier chips, and the bus can be in one of four states, termed idle status Ti, send status Ts, perform command status Tc and hold status Th. Each of these possible states will require one processor clock period. Timing diagrams usually show the **system** clock pulses marked with

NOTE: All times in nanoseconds, ns.

I - Essential timing Requirements (80188 8 MHz)

Symbol	Parameter	80188 Min	80188 Max	80186 Min	80186 Max	80186-10 Min	80186-10 Max	Notes
TDVCL	Set-up time, data in	20		20		15		
TCLDX	Hold time, data in	10		10		8		
TARYHCH	ARDY setup to active state	20		20		15		1
TARDLCL	ARDY inactive set-up time	35		35		25		
TCLARX	ARDY hold time	15		15		15		
TARYCHL	ARDY inactive HOLD time	15		15		15		
TSRYCL	SRDY set-up time	20		20		20		
TCLSRY	SRDY HOLD time	15		15		15		
THVCL	HOLD set-up	25		25		20		1
TINVCH	Set-up INTR,NMI,TEST,TMR IN	25		25		25		1
TINVCL	Set-up DRQ0,DRQ1	25		25		20		1

Notes:
1. To guarantee recognition at next CLK

II - Required Timing Responses

Symbol	Parameter	80188 Min	80188 Max	80186 Min	80186 Max	80186-10 Min	80186-10 Max	Notes
TCLAV	Address valid delay	5	55	5	55	5	44	
TCLAX	Address hold time	10		10		10		
TCLAZ	Address float delay	*	35	*	35	*	30	1
TCHCZ	Command lines float delay		45		45		40	
TCHCV	Command lines valid delay		55		55		45	
TLHLL	ALE width		*		*		*	2
TCHLH	ALE active delay		35		35		30	
TCHLL	ALE inactive delay		35		35		30	
TLLAX	Address HOLD to ALE inactive		*		*		*	3
TCLDV	Data valid delay	10	44	10	44	10	40	
TCLDOX	Data Hold time	10		10		10		
TWHDX	Data hold after WR	*			*		*	4
TCVCTV	Control active delay 1	5	50	5	50	5	40	
TCHCTV	Control active delay 2	10	55	10	55	10	44	
TCVCTX	Control inactive delay	5	55	5	55	5	44	
TCVDEX	DEN# inactive delay (not WR)	10	70	10	70	10	56	
TAZRL	Address float to read active	0		0		0		
TCLRL	RD# active delay	10	70	10	70	10	56	
TCLRH	RD# inactive delay	10	55	10	55	10	44	
TRHAV	RD# inactive-next addr. act.	*		*		*		5
TCLHAV	HLDA valid delay	5	50	5	50	5	40	
TRLRH	RD# width	*		*		*		6
TWLWH	WR# width	*		*		*		7
TAVLL	Address valid to ALE low	*		*		*		8
TCHSV	Status active delay	10	55	10	55	10	45	
TCLSH	Status inactive delay	10	65	10	65	10	50	
TCLTMVC	Timer output delay		60		60		48	
TCLRO	Reset delay		60		60		48	
TCHQSV	Queue status delay		35		35		28	
TCHDX	Status hold time	10		10		10		
TAVCH	Address valid to Clock HIGH	10		10		10		
TCLLV	LOCK# valid/invalid delay	5	65	5	65	5	60	
TCLCSV	Chip-select active delay		66		66		45	
TCXCSX	Chip-select HOLD	35		35		35		
TCHCSX	Chip-select inactive delay	5	35	5	35	5	32	

NOTES:
1. Equal to TCLAX time
2. 80188-TCLCL-35; 80186-TCLCL-35; 80186-10 -TCLCL-30
3. 80188-TCHCL-25; 80186- TCHCL-25; 80186-10 -TCHCL-20
4. 80188-TCLCL-40; 80186-TCLCL-40; 80186-10 - TCLCL-34
5. All chips - TCLCL-40
6. 80188 - 2TCLCL-50; 80186- 2TCLCL-50; 80186-10 - 2CLCL-46
7. 80188 - 2TCLCL-40; 80186-2TCLCL-40; 890186-10 - 2TCLCL-34
8. 80188 - TCLCH-25; 80186 - TCLCH-25; 80186-10 - TCLCH-19

* Depends on other timings

Figure 5.20 *Summary of timing requirements for 80188/80186.*

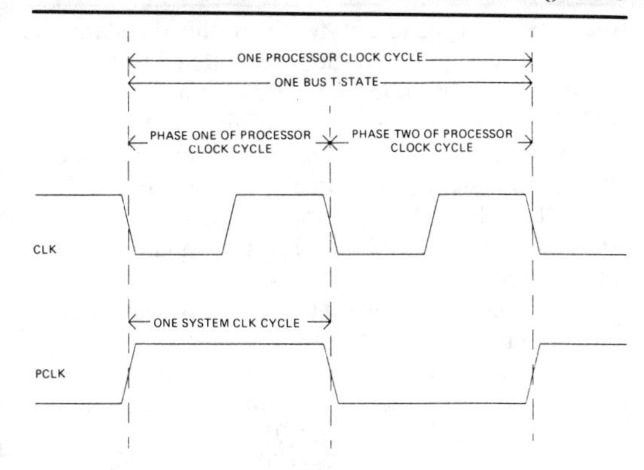

Figure 5.21 *Phases of the 80286 clock cycle.*

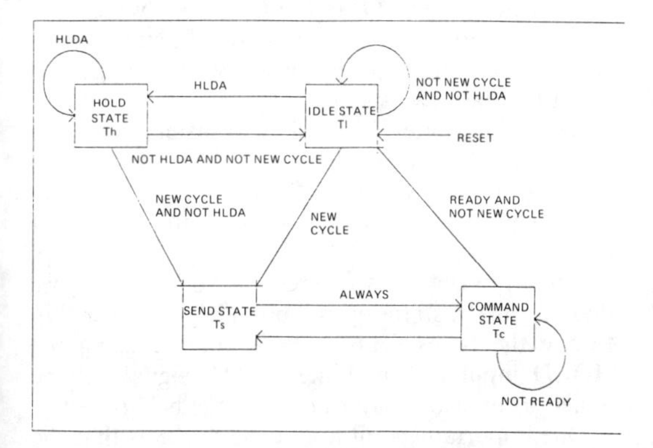

Figure 5.22 *Possible state sequences for the 80286 bus cycles.*

phase numbers for the corresponding processor clock.

The possible sequences of states are illustrated in Figure 5.22. The fastest possible sequence is of command followed by send, then back to command for another cycle. A bus cycle will **always** start with a command state, and this is also the pattern on READY. A send state is **always** followed by a command state. Delays are always implemented in the command state, so that it is possible to move from a command state to an idle state. The idle state is also entered at a RESET, and the processor can be maintained in the idle state. From the idle state, the processor can move to the send state or to the hold state. The hold state can be maintained, or state switched to idle or to send according to the voltages on the various pins. The diagram shows the state of the HLDA signal for transitions that involve hold and idle states. Most 'normal' sequences use a send state, one or more command states, and an idle state in a complete bus cycle.

The differences between the four states are important. In send state, Ts, the address is available on the address pins, and data will be sent out on a write cycle (read cycles receive data in Tc). Status signals are also available so that the usual 82288 bus controller chip can decode the status lines and generate signals for the bus transceivers. The command state is used for data transfers, reading data from memory or I/O and transferring write data also. If memory or I/O is slow, the Tc state can be repeated (determined by the READY voltage), so being used as a wait state.

In the idle state, Ti, no data transfers are being requested or being executed. The change from Ti to Ts is signalled by the S0 or S1 line going LOW, and this also means that the clock state is entering phase 1 (φ1). The buses are **not** floated during an idle state, but if a hold state is entered all address pins are floated and the status output pins allow another chip to use the buses. This state is entered using the HOLD input pin, and the HLDA signal will be issued to indicate that a hold state is being used.

One important point about the 80286 is that the processor itself uses remarkably few control signals. This is because the 80286 is presumed to be used along with support chips such as the 82288 bus controller, which decode the few control outputs of the processor and issue the necessary additional signals. This is why there is no obvious read/write enabling signal, only the set of COD/INTA, M/IO, S1 and S0 which are used as inputs to the 82288 and which provide all the necessary information by decoding.

The fundamental difference in the 80286 timing is that pipelining is possible on the local bus. This allows a new bus operation (reading or writing) to be started after two processor cycles, even if an individual bus action has taken three processor cycles to

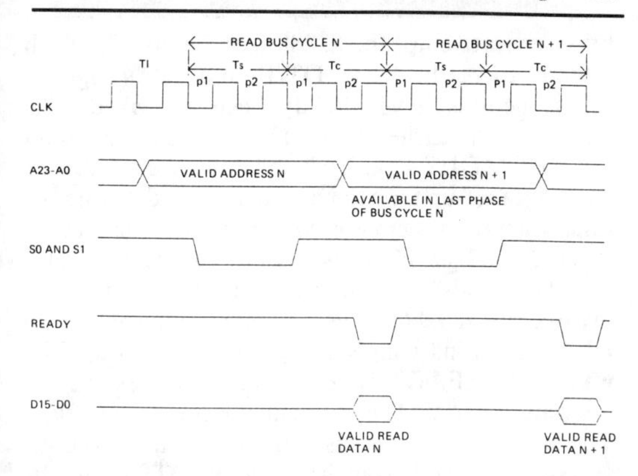

Figure 5.23 *A pair of read bus cycles with pipelining used.*

complete. This is done by allowing the first clock of a bus operation to overlap the last clock of a previous bus action, and requires address decoding and logic to operate ahead of establishing bus voltages. The address for a new bus operation can be put out during the second phase of a Tc cycle, so that external latching is needed if the address is to be held constant throughout an entire Ts–Tc bus cycle. The use of pipelining makes it necessary to consider bus cycles in pairs.

Figures 5.23 shows a pair of read bus cycles in which pipelining is being used. The phases of the states are illustrated by showing the system clock as well as the processor clock, and the address bus is shown as being active before the start of the Ts state, due to a previous pipelining action. The status logic signal S0 AND S1 goes LOW at the start of the Ts state, during which time the address is maintained for this read on the address bus. In the Tc state, the address bus can change in the second phase, and the data bus will be used by the data **for the first address** which has now been read, allowing for the delay in reading memory. The Tc state is terminated by the READY signal, which would have been delayed so as to repeat the Tc state if the memory had been slow, requiring some wait states.

By this time, the next address already exists on the lines, and can be used to ensure that the next memory access is ready. The bus control signals are handled by a separate chip, the 82288 (see Section 7) which attends to ALE, R/W, DT/R and DEN signals, whose actions have already been described with respect to the earlier chips. This chip can also send out the CMDLY signal, which allows the address or write data set-up time to be increased by delaying the time when the system bus command becomes active. Since the 82288 chip also issues the MRDC and MWTC signals for reading and writing, this does not involve the 80286. This offers an alternative to READY for inserting waiting into bus actions. The **wait due** to READY is called a command extension, the wait due to CMDLY is called a command delay. The 82288 chip will also allow the use of both synchronous and asynchronous READY signals, ARDY and SRDY.

Instruction fetch cycle

An instruction fetch will start before the current instruction is executed. This is the **instruction pre-fetch** action, and is possible only if the local bus (between the 80286 and any bus-control or transceiver chips) is not used for data reading or writing. The pre-fetched instructions are stored in a queue in an internal register of the 80286, and a pre-fetch cannot take place unless one word (2 bytes) of the queue is empty. Any pre-fetching will stop if a control transfer (jump) instructor or an HLT instruction has been decoded and placed into the instruction queue.

In addition, there are restrictions on instruction placement in memory. An instruction that is one word long will be pre-fetched irrespective of whether the word starts at an odd or an even address number. A single byte instruction, however, can only be read from an odd-numbered address. In addition, there are safeguards against breaking protection in protected mode. In real (emulating 8086) mode the pre-fetcher can maintain up to 6 bytes (in the absence or HLT or JMP instructions), but in protected mode the pre-fetcher can never operate beyond the end of the code segment. An exception 13 will be generated if the program attempts to execute an instruction following the last instruction in a segment.

The basic bus cycle illustrated in Figure 5.23 shows two adjacent read cycles, and the other main data transferring action, writing, is best described by

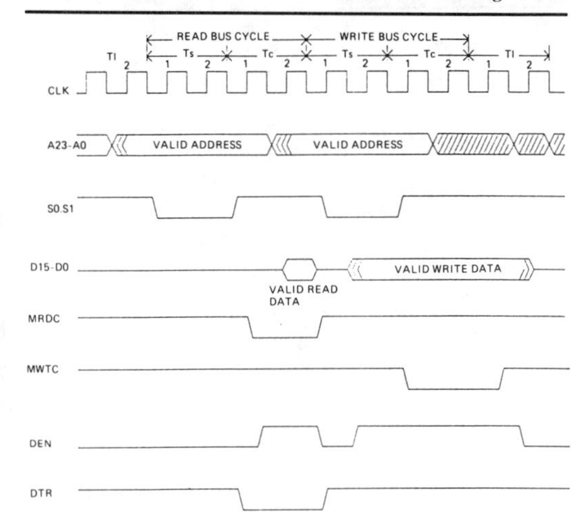

Figure 5.24 *A read–write pair of cycles with pipelining used.*

showing two cycles of read–write, write–read and write–write, with the signals of the 82288 chip shown alongside the 80286 signals to indicate the relative timing. Each cycle is shown as a pair of bus cycles followed by an idle; a typical arrangement.

Figure 5.24 shows the read–write pair of cycles, in which the MRDC, MWTC, DEN and DT/R signals are provided from the 82288 chip. The MRDC signal is the memory read data command which is sent out to memory as a command to place data on to the data bus. The MWTC is the corresponding memory write command, both active LOW. The DEN signal is the usual data enable for transceivers, and DT/R is the data direction (transmit or receive) output. As usual, the address for the first cycle, the read, is established before the Ts part of the cycle, and the start of φ1 of Ts is marked by the S0/S1 pair going low. The start of φ1 of Tc is marked by the MRDC and DTR signals from the 82288 and followed by DEN, so enabling the data on to the data bus. By this time the address has changed to the write address, and data will be put onto the data bus from the 80286 in the Ts state of the write cycle. The 82288 will issue corresponding DTR and DEN levels, and when the MWTC signal is sent from the 82288 in the first

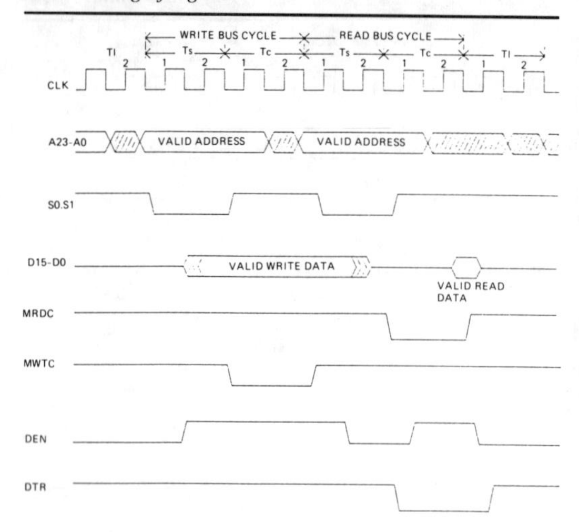

Figure 5.25 *A write–read pair of cycles with idle states leading and trailing.*

phase of Tc, data can be written to the memory. By this time, the address lines no longer have valid information, since the following state is an idle state. Valid addresses are latched by an ALE signal from the 82288.

Figure 5.25 illustrates the write–read cycle followed by an idle state. The address for writing has become valid just before the start of the Ts state, marked as usual by S0/S1 both going LOW. The 82288 issues DEN by the end of the first phase of Ts, so that write data is put on to the bus and has settled before the MWTC strobe enables writing to the memory. There is a short period in the second phase of Tc for which no valid address exists, and then the read address is established just before the start of the Ts of the read cycle. The 82288 sends out appropriate DT/R and DEN signals, and the MRDC strobe allows the data on the bus to be read. In this time, the address lines will no longer hold a valid address because of the following idle state.

Figure 5.26 shows a write–write cycle followed by an idle state. The address for the first write is established before the start of the cycle as usual, with S0 and S1 marking the start of Ts. Data settles on the bus by the second phase of Ts, and in Tc the MWTC

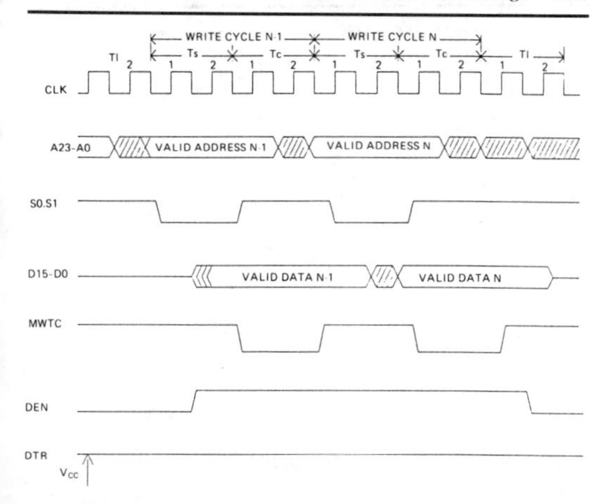

Figure 5.26 *A write–write pair of cycles with idle states leading and trailing.*

signal put out by the 82288 chip will strobe the data to memory. The address changes, and becomes valid for the second write just before the start of the second Ts state. The new data does not settle until about the start of the Tc state, and MWTC signal is once again put out from the 82288 so as to write this second batch of data. The address lines hold invalid numbers from the second phase of this Tc, since the following state is an idle state.

Using HOLD

The use of HOLD and HLDA on the 80286 is illustrated in Figure 5.27, starting with the processor in a Th state, and showing how a subsequent write is affected by a HOLD input. The write cycle is completed, and one idle state is entered so as to ensure that the write data is held for long enough. The Th state is then started, with HLDA signalling the start of the hold. Note that the start and stop of the write action have been triggered by CMDLY and ARDY respectively, using an inactive SRDY pin on the 82288. Note that HOLD must not be used following RESET until 34 system clock pulses have elapsed.

Looking at the diagram in detail, the HOLD signal has been active and is taken low at the end of

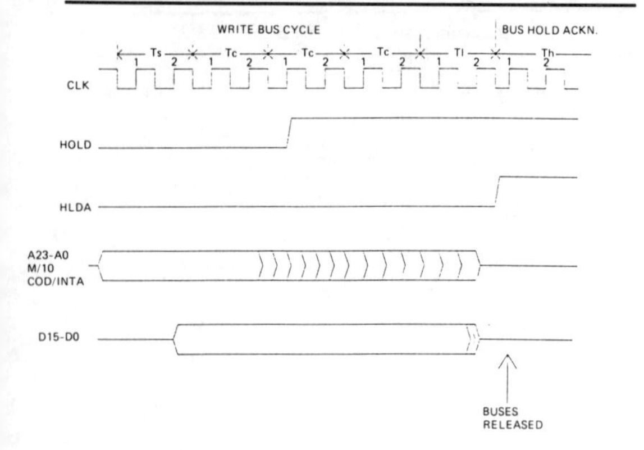

Figure 5.27 *Using HOLD/HLDA* on the 80286, showing the transition from a write bus cycle to acknowledgement of the HOLD. The SRDY or ARDY signal would be used to end the HOLD.

the first phase of the first of the Th states shown here. The HLDA signal follows at the end of the second Th, the minimum time possible between HOLD and HLDA. One more Th follows, in which the address bus establishes a valid address, and signals BHE and LOCK also have valid values. The write cycle then starts, using one Ts state, and three Tc states.

In the second Tc state, HOLD has gone HIGH again, and this causes HLDA to go high at the end of the idle state that follows the second Tc. Once again, this represents the minimum HOLD to HLDA time. The data on the write bus is valid into the idle state. The write has been delayed by the use of CMDLY, and terminated by #ARDY from the 82284 chip.

Interrupt sequences take the form indicated by Figure 5.28. As usual, this requires two INTA cycles, with INTA generated from the 82288 chip, and each INTA cycle needs at least one wait state (extra Tc state) so as to make correct use of the 82C59A-2 programmable interrupt controller chip, which needs about 100 ns. The LOCK signal is activated in the first INTA cycle to prevent the 82288 chip from releasing the buses between INTA cycles. At the end of the second INTA cycle, the vector address appears on D0–D7 of the data bus, with the higher-order lines ignored. Figure 5.29 is a summary of 80286 timing requirements.

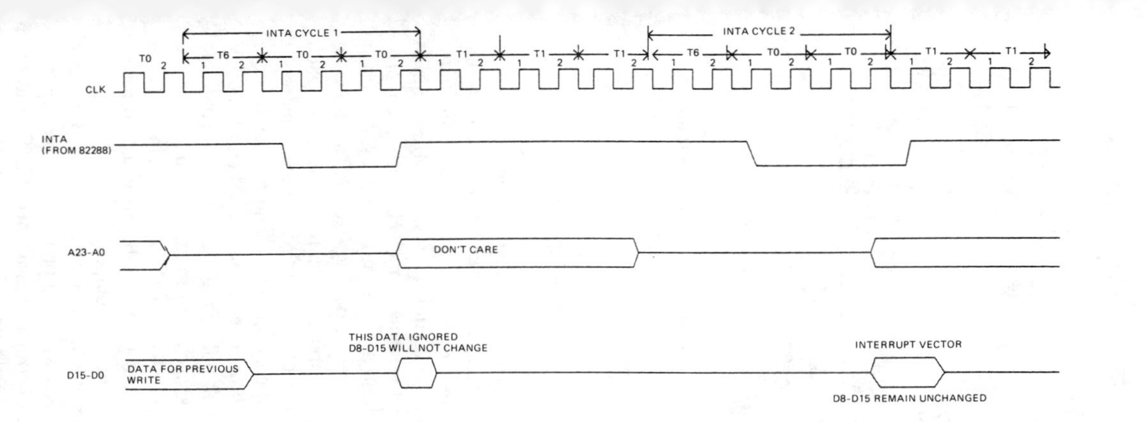

Figure 5.28 *The interrupt sequence of the 80286.*

NOTE: All times in nanoseconds, ns.

Parameter	8 MHz Min	8 MHz Max	10 MHz Min	10 MHz Max	12.5 MHz Min	12.5 MHz Max	Notes
Asynch. inputs setup time	20		20		15		1
Asynch. inputs Hold time	20		20		15		1
RESET setup time	28		23		18		
RESET Hold time	5		5		5		
Read data setup time	10		8		5		
Read data hold time	8		8		6		
READY# setup time	38		26		22		
READY# Hold time	25		25		20		
Status/PEACK# valid delay	1	40	--	--	--	--	2,3
Status active delay	--	--	1	22	3	18	2,3
PEACK# active delay	--	--	1	22	3	20	2,3
Status/PEACK# inactive delay	--	--	1	30	3	22	2,3
Address valid delay	1	60	1	35	1	32	2,3
Write data valid delay	0	50	0	30	0	30	2,3
Address/Status/Data float delay	0	50	0	47	0	32	2,4,7
HLDA valid delay	0	50	0	47	0	27	2,3
Address valid-status valid setup	38		27		22		3,5,6

Notes:

1. INTR,NMI,HOLD,PEREQ,ERROR#,BUSY# inputs
2. Delay from 1.0 V on CLK to 0.8V or 2.0V or float
3. Load capacitance 200 pF
4. Floats when output current is less than I_{LO}
5. Measured from address 0.8V or 2.0V to 0.8V or 2.0V status level
6. For load capacitance of 10pF or more, subtract 7ns.
7. Not tested.

Figure 5.29 *A summary of the timing requirements of the 80286.*

The 80386DX and 80386SX timings

The timing pattern of the 80386 follows very closely the patterns established by the 80286, but with one important change. As before, the clock input frequency is divided by 2, and the resulting frequency is referred to as the **processor clock**, with each processor clock cycle divided into two **phases**, one for each complete system clock cycle, CLK2 as shown in Figure 5.30. An important point to note is that each transition of the processor clock takes place on a **rising** edge of the system clock and not, as is more usual, on a falling edge. Timing is taken at the 2 V level of the clock pulse, and the time for a complete processor clock cycle can range from 62 ns for the 16 MHz clock to 40 ns for the 25 MHz clock. These clock pulses would normally be provided from the 82C284 clock generator chip.

As before, a bus cycle consists of two processor clock cycles, each with two phases, and the bus can be in any one of four states, termed T1, T2 and the idle start Ti and hold state Th. Each of these possible states will require one processor clock period. Timing diagrams usually show the **system** clock pulses

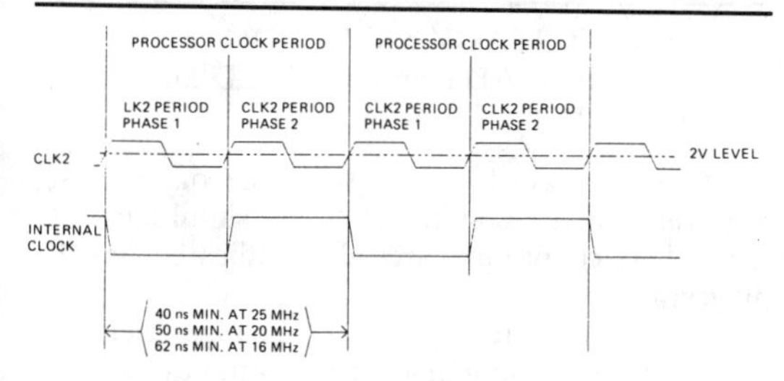

Figure 5.30 *The processor clock cycles and its phases for the 80386.*

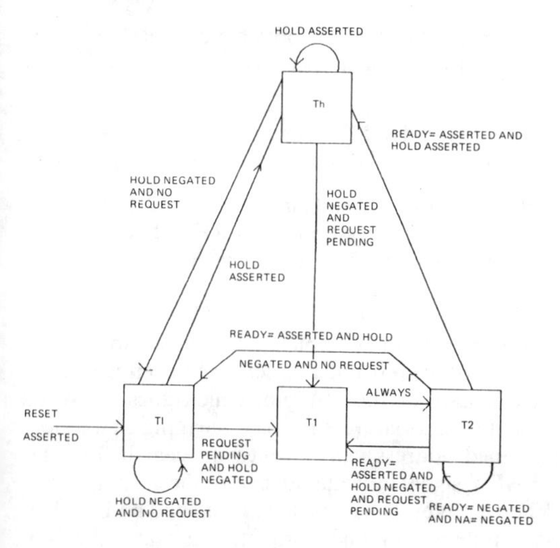

Figure 5.31 *The possible states and sequences for non-pipelined bus actions.*

marked with phase numbers for the corresponding processor clock.

The possible sequences of states depend on whether pipelined bus actions are being used or not. The sequence diagram of Figure 5.31 is for addressing that is **not** pipelined. The T1 state will be followed immediately by the T2 state. The T2 state may be repeated, acting as a wait state, or one of the following paths may be taken:

- to T1, if READY and no HOLD and a request pending;

- to Th, if READY and HOLD;
- to Ti, if READY and no HOLD and no bus request;

– note that the word 'and' is used in the logical sense, requiring all the conditions to be simultaneously true. The conditions are also indicated in the diagram.

From the Th state, the possible changes include a repeat of the Th state if HOLD continues, or

- to T1, if no HOLD and bus request pending;
- to Ti, if no HOLD and no bus request.

From the Ti state, the possible changes include a repeat of the Ti state if there is no HOLD and no bus request, or

- to Th if HOLD is used;
- to T1 if no HOLD and bus request pending.

A RESET always results in entering the Ti state, so that processing will start after a RESET when a bus request is awaiting attention, provided that HOLD has not been used.

The state table for a pipelined addressing system is considerably more complicated, because not all bus actions will be pipelined. Pipelining depends heavily on the state of the NA pin, which goes LOW to indicate a next address request, meaning that a new pipelined address can be put out even before the READY signal indicates an end to the current cycle. When a bus cycle has started and a valid address has been put out for one complete bus state, the NA signal is sampled at the end of every phase 1 until the bus cycle is acknowledged.

The diagram (Figure 5.32), shows the same four basic states with the same paths between them, but three extra states have been added. These are:

- T1P, the first clock of a pipelined cycle;
- T2l, the later part of a pipelined cycle when NA is LOW but no bus request is available;
- T2P, a subsequent cycle of pipelining when NA is LOW and there is a bus request pending.

Many of the transitions that are shown in this latter diagram are seldom used. If NA is held LOW, for example, only the states T1P and T2P will be used, with *all* addressing pipelined. If NA is held high, the

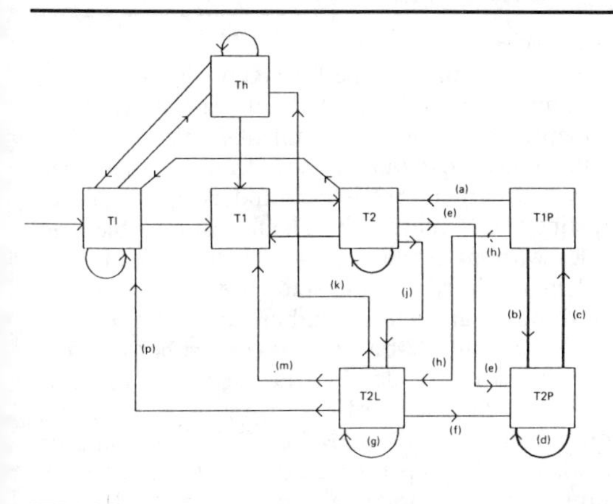

TRANSITION CONDITIONS
(a) NA# NEGATED
(b) NA# ASSERTED AND HOLD NEGATED AND REQUEST PENDING
(c) READY# ASSERTED
(d) READY# NEGATED
(e) READY# NEGATED AND NA# ASSERTED AND HOLD NEGATED AND REQUEST PENDING
(f) READY# NEGATED AND REQUEST PENDING AND HOLD NEGATED
(g) READY# NEGATED AND (NO REQUEST OR HOLD ASSERTED)
(h) NA# ASSERTED AND (HOLD ASSERTED OR NO REQUEST)
(j) (NO REQUEST OR HOLD ASSERTED) AND (NA# ASSERTED AND READY# NEGATED)
(k) READY# ASSERTED AND HOLD ASSERTED
(m) READY# ASSERTED AND HOLD NEGATED AND REQUEST PENDING
(p) READY# ASSERTED AND HOLD NEGATED AND NO REQUEST

Figure 5.32 *The possible states and sequences for pipelined bus actions.*

state diagram reverts to its non-pipelined version. Note that there is **no path** to Ti or Th states from T1P or T2P, and such states can be entered only by a transition from T1P to T2, reverting to non-pipelined operation, or by way of T2l if the NA signal goes high in this state. The path for a change from non-pipelined addressing to pipelined addressing is by way of T1, T2 and T2P; a path back from pipelined operation to non-pipelined operation makes use of T2 or T2l.

The active states of the processor consist of any read or write involving memory or ports, the interrupt acknowledge, and the halt or shutdown indication. When none of these actions is being carried out, the processor must be in the Ti state or the Th state. The Ti state is signalled by the address strobe output ADS remaining high beyond the end of one bus cycle, and the Th state is signalled by the HLDA signal output.

Read cycles

Read cycles may be performed with or without pipelining, and with or without wait states – the principle of interleaving banks of relatively slow memory so as to avoid the use of wait states has been described earlier (Section 2) and is applicable to pipelined addressing only. The fastest possible read cycles without pipelining are illustrated in Figure 5.33, in which the only bus states used are T1 and T2. Valid bus signals are established in φ1 of T1, providing a full 32-bit address (remember that the lower bits are provided by BE0–BE3 so as to allow the use of double-word, word, or byte reads and address alignment). The M/IO, D/C, W/R outputs are also valid at this time. In φ2 of T2, the READY output goes LOW to allow data to be read at the end of T2 (the data is still available at the start of the next T1). As for the 80286, the READY signal indicates the end of a bus cycle, and T2 states will continue until READY is put LOW. This fast bus cycle is possible for consecutive reads only if the memory is fast enough (including the use of cache memory). In these non-pipelined bus actions, the address and the bus definition bits remain steady throughout the bus cycle.

The pipelined equivalent read cycles are illustrated in Figure 5.34. In this example, the address for the first read is already valid, due to the pipelining, at the start of T1P. The address status signal ADS is going

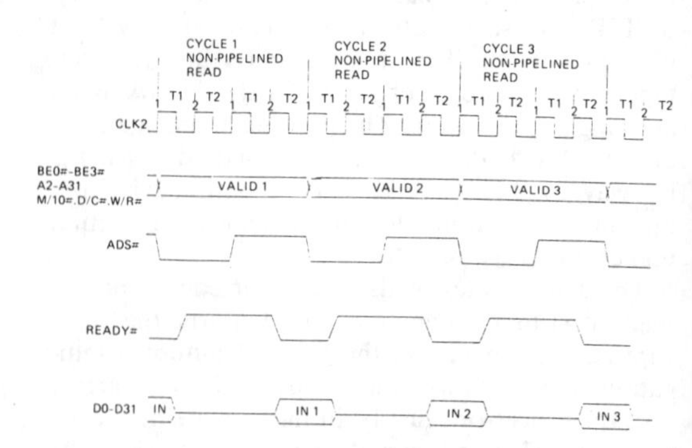

Figure 5.33 *Illustrating the fastest possible read cycle pair with no pipelining. Note that these use states T1 and T2 only.*

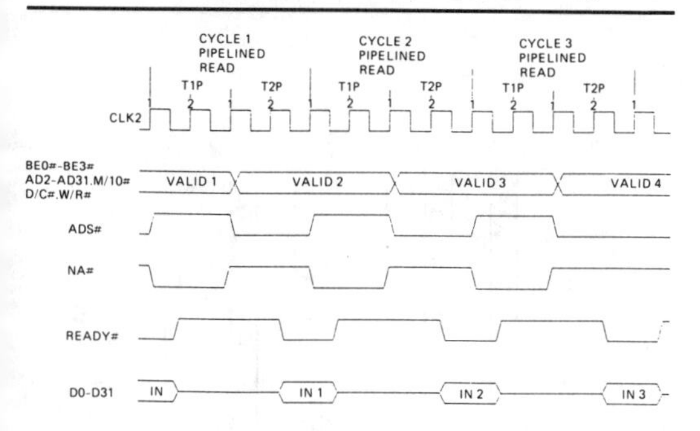

Figure 5.34 *The fastest possible read cycles with pipelining used. Note that these use states T1P and T2P only.*

high in this state to prevent any use of the address when the change is made, and the NA signal goes LOW to force the next address to be used for pipelining in φ1 of T2P. The READY signal goes low in φ2 of T2P in order to read the first batch of data. From then on, the address that is being put out and strobed by ADS in T2P is the address for the **subsequent** read that will take place in the φ2 of the T2P of the **next** cycle. The use of ADS is therefore an essential part of pipelining.

Since an address is available, using pipelining, one state ahead of the time when data is read, the memory has more time in which to respond when pipelining is used than when pipelining is not used. If, for example, the memory requires one wait state for non-pipelined addressing, then no wait states are required for pipelined use. The advantage of pipelining becomes particularly noticeable when memory bank interleaving is used. Note that the use of the NA signal input allows pipelining to be selected on a cycle-by-cycle basis, so that some bus actions can be pipelined and others not. The use of bus hold will always have the effect of breaking out of pipelining.

Read and write, non-pipelined

Figure 5.35 illustrates various combinations of read and write, with idle states – these are for illustration only and do not imply that idle states have to be inserted. In these non-pipelined cycles, address numbers and definition signals (M/IO, D/C, W/R) are

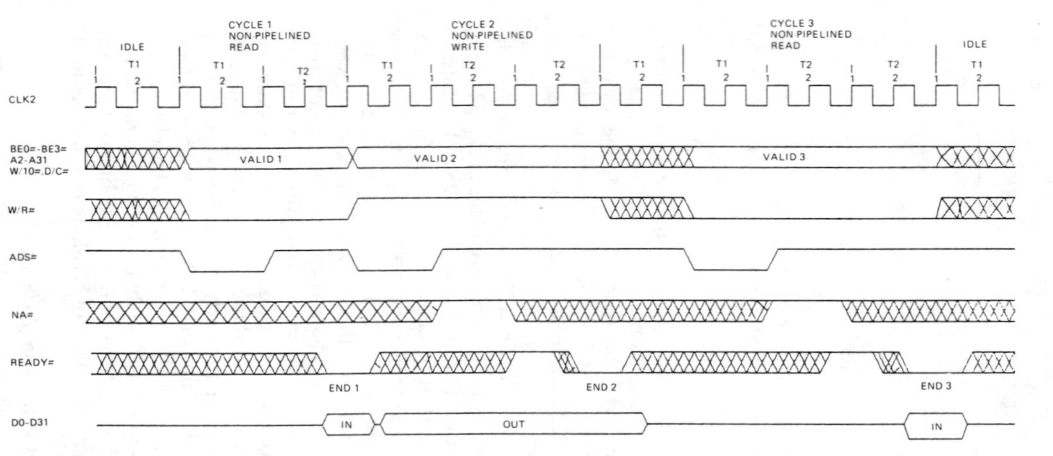

Figure 5.35 *Combinations of read and write cycles with idle and wait states, non-pipelined. These are shown only to illustrate transitions; a write cycle is not necessarily followed by an idle state. Note that data must be valid before READY# is asserted.*

valid throughout the cycle, but the address strobe ADS goes low only in the T1 period to allow the signals to be externally latched. The W/R signal is high for write, low for read, and the READY signal, as usual, goes LOW to indicate the end of a bus cycle – the timing of this signal determines the use of any wait states. Note that the idle state is entered because ADS has not gone low at the start of a T1 state. During the Ti state, the numbers on the buses are of no interest and are not valid.

In the read cycle, data is placed on the (floating) data bus at the end of the T2 state. The memory must be organized in such a way that the data is fully valid when the READY signal goes LOW. On the write cycle, data is put on to the lines in φ2 of T1 and will remain there until φ1 of the state following the acknowledgement by READY going LOW. To maintain non-pipelined action, NA can be held high throughout, but an alternative is to ensure that NA is high in the wait T2 states – any T2 state other than the last in a bus cycle.

During a set of cycles as illustrated here, the BS16 pin of the 80386DX can be used to switch the bus size between 16-bit and 32-bit use, and the LOCK output will appear when the buses must be retained by the processor and not surrendered to other chips such as the 80387 co-processor. At the start of any bus cycle, the data bus will be 32 bits wide, and BS16 is sampled before the READY signal goes low so that the bus width will be determined before the actual read or write of data. When BS16 has been used to make a 32-bit data fetch use two 16-bit cycles, the BS16 signal must be low in the second cycle also, because using one 16-bit bus cycle does not enforce this size on a second cycle. The 80386SX has no BS16 pin, and cannot use a 32-bit bus, so that these comments are not applicable.

Read and write – pipelined

Since the 80386 always starts up following a RESET in non-pipelined states, the change from non-pipelined actions to pipelined actions is important. One odd feature of the 80386DX is that if BS16 and NA are both LOW, then BS16 has priority, and the data bus will be used as a 16-bit bus, but pipelining will **not** start. This does not, of course, apply to the 80386SX. When pipelining has to be started on the

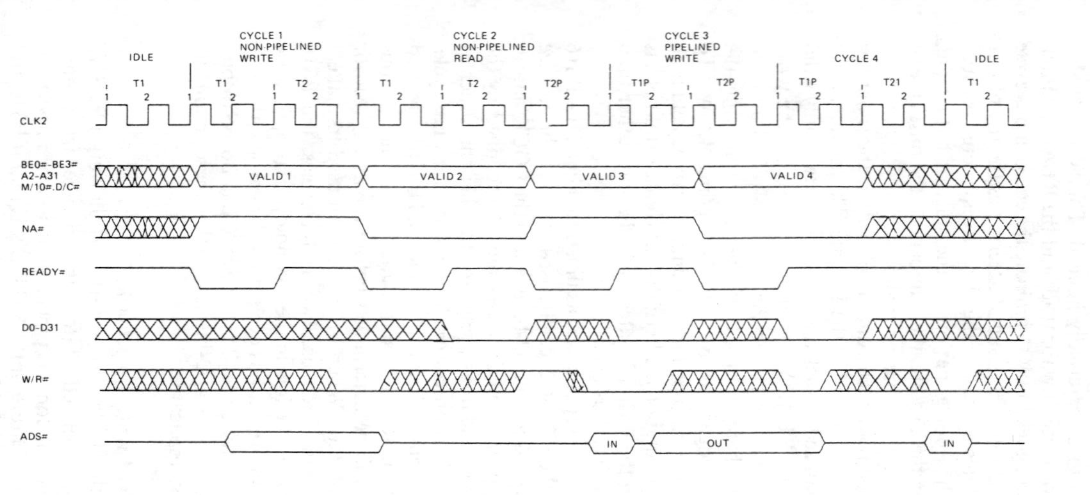

Figure 5.36 *Making the transition from non-pipelined to pipelined actions.*

80386DX, BS16 must be HIGH, enforcing a 32-bit bus, in the T2 state of the last non-pipelined cycle and on each T1P of each pipelined cycle. This allows the BS16 pin voltage to be lowered before the READY signal if needed in order to make use of a 16-bit bus on the 80386DX.

Figure 5.36 shows a typical transition from non-pipelined actions to pipelined actions in a set of read–write actions. The switch to pipelining is carried out, as usual, by making the NA voltage LOW, but this takes effect only when sampled, and NA is sampled only during the wait states of non-pipelined actions. A change from non-pipelined to pipelined action will therefore be delayed after NA has been placed LOW until a wait state appears.

The diagram shows a non-pipelined write action followed by a non-pipelined read. The NA input is taken LOW during T2 of the non-pipelined read and because READY is held HIGH at the end of this T2, the processor enters a T2P wait state. In this state, a new pipelined address becomes valid for the data of the third cycle, the pipelined write, and from that point on, the sequence is of T1P and T2P states, maintaining the pipelining so long as NA is taken low in each T1P part of the cycle. The BS16 signal is not shown here, for the sake of clarity, and for the 80386DX chip, the BS16 signal would be high at each point where NA was sampled.

Any idle state automatically enforces a return to non-pipeline actions, and the fastest change from an idle state to pipelining is achieved by following the idle state with a non-pipelined read or write which incorporates one wait state, with the NA voltage taken low in the wait state. Once again, the 80386DX would require the BS16 voltage to be HIGH during the sampling period of the NA signal. The following other rules apply:

- After NA is taken low, a pending internal bus request will cause the next state to be T2P. If there is no pending bus request, the next state will be T2l. The appearance of the pipelined address is signalled by ADS going low.
- If a bus cycle of the 80386DX **must** use 16-bit data, so that BS16 will be LOW, then NA should not be put low in that bus cycle. Pipelined addressing on the 80386DX is **not**

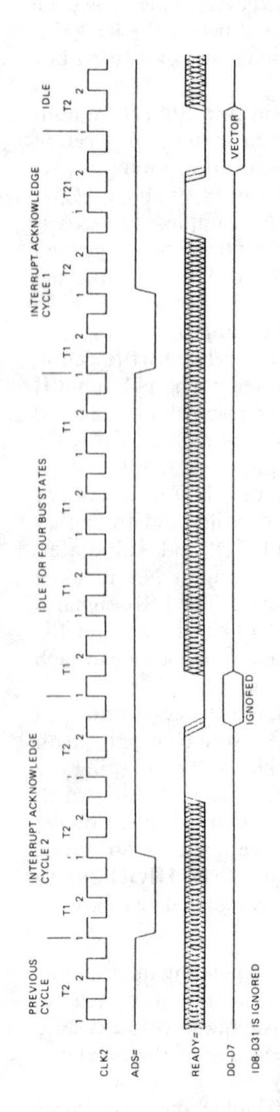

Figure 5.37 *Interrupt timing for the 80386 showing the two INTA cycles.*

compatible with using the 16-line data bus to read 32-bit numbers, since the same base address must be held in A2–A31 for both 16-bit cycles. This conflict does not arise in the 80386SX in which the addresses change in 16-bit steps rather than 32-bit steps.

- An address that is indicated as being valid by ADS going LOW will remain on the bus for at least two processor states. Addresses cannot be changed more rapidly than every two states.
- Pipelining cannot extend to more than one bus cycle ahead.

Interrupt cycles

When the INTR input receives an interrupt request, assuming that interrupts are enabled at the time, the 80386 will carry out two interrupt acknowledge (INTA) cycles. As for the 80286, the vector address for the interrupt handling routine will appear on the D0–D7 lines on the second phase of T2 or T2l of the second interrupt acknowledge. The use of NA is ignored during INTA cycles, because each interrupt acknowledge is followed by idle states. During each INTA cycle, the address lines A3–A31 of the 80386DX are **all low**, and line A2 is used to distinguish the second INTA cycle from the first.

The timing is illustrated in Figure 5.37. The ADS voltage goes LOW in T1 of each INTA, though the address on the lines for the first INTA is 4 and for the second INTA is 0. Any data on the data bus at the end of the first INTA cycle is ignored, as also is data on the higher order lines when the vector number is available on D0–D7. The diagram shows four idle states between the two INTA cycles, in which time the content of the address bus, NA, BS16 and READY will be ignored. This set of four idle states is inserted automatically, and the end of this set is indicated by ADS going low. The data bus floats during T1 and the first T2 of each INTA cycle.

Halt and shutdown

A processor HALT is enforced by the software HALT instruction, and is ended when an interrupt or a RESET signal is received. A processor shutdown is caused by the internal protection system responding to a double fault (a fault encountered when processing another fault). In both conditions, M/IO is

HIGH, D/C is LOW, W/R is HIGH and address numbers A2–A31 are LOW. On the 80386SX the address numbers A2–A23 are low. The distinction is carried out by the BE pins of the 80386DX, which provide the address 2 for a HALT and address 0 for a shutdown. For the 80386SX, the distinction is made using the A1 pin to provide addresses 0 or 2. The HALT must be acknowledged by the READY input being taken LOW, and will be followed by idle states until an interrupt or a RESET restarts the processor. A HOLD input can be entered during this time.

On a shutdown, the cycle is very similar and only the address indicates which form of action is proceeding. After a shutdown the processor cannot be restarted by an INTR form of interrupt, only by the NMI or the RESET. Like a HALT cycle, the shutdown cycle must be acknowledged by READY going LOW, after which the processor will start its run of idle states.

Hold and reset

The hold state Th is used as a response to the HOLD input, with HLDA used to acknowledge the hold. The timing is illustrated in Figure 5.38, showing HOLD going active (HIGH) during an idle state, and HLDA responding by going HIGH on the next state, which is the first Th state. HLDA is the only output with a definite signal during a HOLD; all

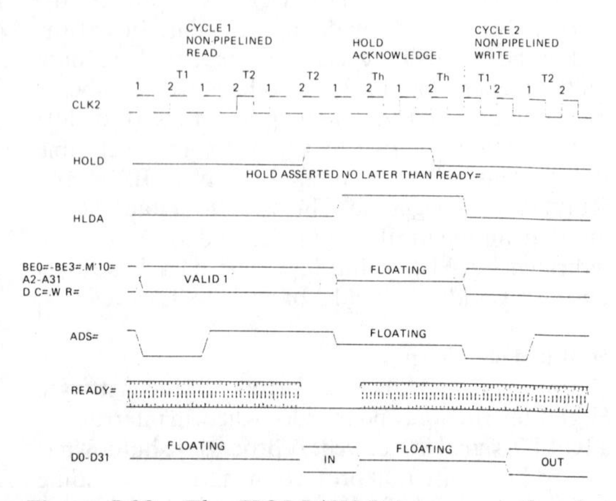

Figure 5.38 *The HOLD/HLDA sequence for the 80386.*

other output and bidirectional signal pins are floated. The only active inputs are HOLD, RESET, BUSY, ERROR and PEREQ, though one NMI can be stored for attention following the HOLD period. Note that there are no internal pull-up resistors on outputs of the 80386, so that if the outputs that need to remain high for no action (like ADS) are to be pulled up, external resistors should be used.

A HOLD can be entered from an idle, as shown or after another cycle has been acknowledged by READY going LOW – in this case, the HOLD voltage must go high before or coincident with the READY going low. HOLD cannot be acknowledged during a LOCK cycle, and if HOLD is applied during the LOCK period, an intermediate cycle may be needed after LOCK is released in order to implement the HOLD action. On the 80386DX, if a HOLD occurs while the BS16 signal is being used to enforce 16-bit data bus use, the acknowledgement of the HOLD will be delayed until the second 16-bit bus cycle has been completed.

A HOLD ends when the HOLD input is taken low again. If no bus request is pending, an idle state will follow the end of the HOLD, but if a bus request is pending, the next state will be T1 (note that a pipelined state cannot follow a HOLD).

The RESET signal has the highest priority among input signals and can interrupt any other processor activity, causing a bus cycle to be abandoned or any idle or bus hold states to be changed. RESET therefore takes priority over HOLD, so that an existing Th state will be abandoned when RESET is used. A RESET signal must remain active for at least 15 CLK2 periods to be recognized, and at least 78 CLK2 periods if the self-test action (see later) is to be carried out following RESET.

The effect of RESET is to drive the pins into specified conditions, as shown earlier in Figure 4.27, with the data bus floated. At the end of a RESET, voltages on some pins can be sampled to indicate actions to be carried out. The BUSY input will be sampled, and if active (LOW) will cause a self-test of the processor to be carried out. The ERROR pin is also sampled, and the result of this sampling will be used to check for the presence of an 80387 co-processor or the alternative, which is the 80287 or none.

NOTE: All times in nanoseconds, ns. Min. operating frequency is 4 MHz.
Parameters for 25 MHz use are preliminary.

Parameter	16 MHz Min	16 MHz Max	20 MHz Min	20 MHz Max	25 MHz Min	25 MHz Max	Notes
CLK2 period	31	125	25	125	20	125	
CLK2 high time	9		8		7		1
CLK2 low time	9		8		7		2
CLK2 fall and rise times		8		8		7	
A2-A31 valid delay	4	36	4	30	4	21	
A2-A31 float delay	4	40	4	32	4	30	
BE0#-BE3#, LOCK# valid delay	4	36	4	30	4	24	3
BE0#-BE3#,LOCK# float delay	4	40	4	32	4	30	
W/R#,M/IO#,D/C#,ADS# valid delay	6	33	6	28	4	21	
W/R#,M/IO#,D/C#,ADS# float delay	6	35	6	30	4	30	
D0-D31 write data valid delay	4	48	4	38	7	27	4
D0-D31 float delay	4	35	4	27	4	22	
HLDA valid delay	6	33	6	28	4	22	
NA# set-up time	11		9		7		
NA# hold time	14		14		3		
BS16# set-up time	13		13		7		
BS16# hold time	21		21		3		
READY# set-up time	21		12		9		
READY# hold time	4		4		4		
D0-D31 read set-up time	11		11		7		
D0-D31 read hold time	6		6		5		
HOLD set-up time	26		17		15		
HOLD hold time	5		5		3		
RESET set-up time	13		12		10		
RESET hold time	5		4		3		
NMI,INTR set-up time	16		16		6		
NMI,INTR Hold time	16		16		6		
PEREQ,ERROR#,BUSY# set-up time	16		14		6		
PEREQ, ERROR#,BUSY# Hold time	5		5		5		

Notes:
1. At 2.0 V
2. At 0.8 V
3. Max. LOCK# valid delay for 25 MHz is 21 ns
4. D0-D31 Write data hold time is 2 ns min. at 25 MHz

Figure 5.39 *A summary of the timing requirements for the 80386DX.*

NOTE: All times in nanoseconds, ns. Min. operating frequency is 4 MHz.
Preliminary figures only - see Intel for design information.

Parameter	16 MHz Min	16 MHz Max
CLK2 period	31	125
CLK2 high time	9	
CLK2 low time	9	
CLK2 fall and rise times		8
A2-A31 valid delay	4	36
A2-A31 float delay	4	40
BHE#,BLE#, LOCK# valid delay	4	36
BHE#,BLE#,LOCK# float delay	4	40
W/R#,M/IO#,D/C#,ADS# valid delay	6	33
W/R#,M/IO#,D/C#,ADS# float delay	6	35
D0-D31 write data valid delay	4	40
D0-D31 float delay	4	35
HLDA valid delay	6	33
NA# set-up time	5	
NA# hold time	21	
READY# set-up time	19	
READY# hold time	4	
D0-D31 read set-up time	9	
D0-D31 read hold time	6	
HOLD set-up time	26	
HOLD hold time	5	
RESET set-up time	13	
RESET hold time	4	
NMI,INTR set-up time	16	
NMI,INTR hold time	16	
PEREQ,ERROR#,BUSY# set-up time	16	
PEREQ, ERROR#,BUSY# hold time	5	

Notes:
1. At 2.0 V
2. At 0.8 V

Figure 5.40 *A summary of the timing requirements for the 80386SX.*

During the RESET period the processor goes through an initialization sequence which requires from 350 to 420 CLK2 cycles. If self-test has been requested, this will be carried out, requiring more than 1 000 000 CLK2 cycles and placing the number 00000000H into the EAX register if all of the internal self-test checks have been satisfactorily completed. Any non-zero number indicates a fault in the processor.

Figure 5.39 is a summary of the timing requirements of the 80286DX, and Figure 5.40 is a similar summary for the 80386SX chip.

The i486 timings

Timings for the i486 were still provisional at the time of writing, and are summarized in Figure 5.41. The bus state diagram for the i486 (Figure 5.42), is rather different from that of the 80386 chips, with the main states marked as T1 and T2 – note that the use of an internal cache memory implies that pipelining is not required.

NOTE: All times in nanoseconds, ns. Min. operating frequency is 8 MHz.
Parameters are all preliminary- see Intel for design information.

Parameter	25 MHz		33 MHz		Notes
	Min	Max	Min	Max	
CLK period	40	125	25	125	
CLK high time	14		8		1
CLK low time	14		8		2
CLK fall and rise times		4		8	
A2-A31 valid delay	3	22	4	30	3
A2-A31 float delay	0	--	4	32	3
PCHK# Valid delay	3	27	4	30	
BLAST#, PLOCK# valid delay	3	27	4	32	
BLAST#, PLOCK# float delay	0		0		
D0-D31, DP0-3 write data valid delay	3	22	4	38	
D0-D31,DP0-3 float delay	0		4	27	
EADS# set-up time	8		6	28	
EADS# hold time	3		3		
KEN#,BS16#,BS8# set-up time	8		6		
KEN#,BS16#,BS8# hold time	3		3		
RDY#, BRDY# set-up time	8		6		
RDY#,BRDY# hold time	3		3		
D0-D31,DP0-3,A4-A31 read set-up time	5		5		
D0-D31,DP0-3,A4-A31 read hold time	3		3		
HOLD,AHOLD,BOFF# set-up time	10		8		
HOLD,AHOLD,BOFF# hold time	3		3		
RESET Etc. set-up time	10		8		4
RESET Etc. hold time	3		3		4

Notes:

1. At 2.0 V
2. At 0.8 V
3. Also PWT, PCD, BE0-BE3, M/IO#, D/C#, W/R#, ADS#, LOCK#, FERR#, BREQ, HLDA.
4. Also FLUSH#, A20M#, NMI, INTR, IGNNE#

Figure 5.41 *A summary of the timing requirements for the i486 (provisional).*

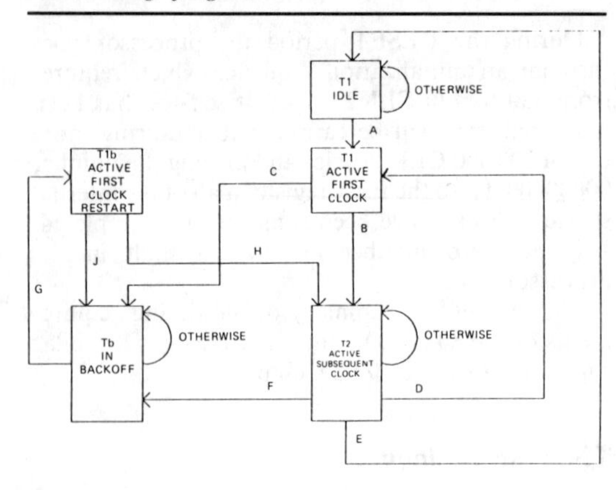

CONDITIONS FOR CHANGES
A NCPEN AND MHOLD# AND AHOLD# AND BOFF#
B BOFF#
C BOFF
D (RDY OR BRDY AND (LASTC OR BABLE#)) AND NCPEN AND MHOLD# AND AHOLD# AND BOFF#
E (RDY OR BRDY AND (LASTC OR BABLE#)) AND (NCPEN# OR MHOLD OR AHOLD) AND BOFF#
F BOFF
G MHOLD# AND AHOLD# AND BOFF#
H BOFF#
J BOFF

Figure 5.42 *The bus states of the i486 – note that pipelining is not used because of the use of built-in cache memory.*

6 Bus structure and co-processor use

8088 bus

Address and data

The 8088 bus set uses a simple unidirectional bus for lines A8–A15, with multiplexing used for AD0–AD7 and for A16–A19. A system using the 8088 can opt to use the multiplexed buses for interconnection, or to demultiplex the 8088 buses and used the demultiplexed form for interconnections. Examples of both bus formats are shown in Section 12.

A bus cycle consists of at least four CLK cycles, T1, T2, T3 and T4, all of the same duration. Each cycle is taken as starting and ending on a negative transition of the system clock as generated by the clock-chip. The address is put out during T1, and data is transferred in T3 and T4, leaving T2 as a changeover period for the multiplexed buses. In addition to these four main states, the T3 and T4 states can be separated by Tw wait states of the same duration, and these will appear if the device (memory or port) that is being addressed is not ready to release the buses. In addition, idle states Ti can be inserted between complete bus cycles when required to allow the processor to carry out internal actions.

In the minimum mode, the 8088 can be used with either the multiplexed bus or the demultiplexed bus. Using the multiplexed bus allows the use of the multiplexed bus peripherals 8155, 8156, 8355, 8755A and 8185, allowing the construction of computers or controllers with a low chip count. A demultiplexed bus can use one latch for 64K addressing or two latches for full 1 Mb addressing, with a third latch used for buffering if required. The latches are controlled by the ALE signal. A transceiver for data can be controlled by the DEN and DT/R signals from the 8088.

In maximum mode, the 82C88 bus controller would be used along with address latching and the data transceiver. This results in a full control bus becoming available (see later, Section 7). The use of maximum mode is essential if a co-processor is to be used.

Control bus

The control bus in minimum mode consists of the ALE, RD, WR, IO/M, INTA, DT/R and DEN signals of the 8088, plus any RESET signal such as can be obtained from the 82C44A clock chip. The bus arbitration signals such as HOLD and HLDA are also control signals, and are dealt with separately. The ALE signal goes high at the clock low of the T1 part of any bus cycle, and is used to latch the address bits. This output is never floated. The RD read strobe goes active LOW on a memory or port read cycle on the local bus. The signal is active during the latter part of T2, T3 and any wait states, and is guaranteed to remain HIGH in T2 until the local bus has been floated.

The WR signal is the write strobe, active LOW in T2, T3 and Tw of any write cycle. The pin is floated during a HOLD. The IO/M control signals allows the choice of memory access (LOW) or port access (HIGH), and its voltage is valid from the T4 **preceding** a bus cycle to the final T4 of the cycle in which memory or port access is used. The INTA signal input is the maskable interrupt.

The DT/R signal is used to control a data transceiver, using HIGH to transmit and LOW to receive data. The timing follows that of IO/M. The DEN data enable signal is a chip enable for a data transceiver which allows the transceiver to separate the data signals from the address signals on the multiplexed lines. DEN goes active LOW during each memory of port access and each INTA cycle. On the read or INTA cycles DEN is LOW from the middle of T2 until the middle of T4. For a write cycle, DEN is active from the start of T2 to the end of T4, and it floats during a HOLD.

The control bus in maximum mode is obtained mainly from the 82C88 bus controller chip, and consists of DT/R, ALE, DEN and INTA signals which perform the same tasks as on the minimum set-up, along with MRDC, MWTC, AMWC, IORC, IOWC, AIOWC and MCE/PDEN signals. All of these are outputs, mostly active LOW. The corresponding inputs to the 82C88 in maximum mode are the S0, S1 and S2 status bits, plus the CLK input and the control signals AEN, CEN and IOB.

The MRDC output is a memory read strobe which goes LOW in the T2 and T3 states of a read operation. MWTC is the corresponding write strobe which goes LOW in the T3 cycle to cause memory to be enabled for writing. For slow memory, the AMWC signal can be used as an advanced warning of a memory write, starting in T2 instead of in T3. The IORC, IOWC and AIOWC outputs perform the same strobing action for in/out port access as the MRDC, MWTC and AMWC set perform for memory.

The MCE/PDEN pin action depends on the voltage level of the IOB input pin. With IOB low, the MCE (master cascade enable) output is active HIGH in an interrupt sequence, and enables the reading of a cascade address from a priority interrupt controller (8259A) to the data bus. Cascading implies that more than one PIC will be used, identified by a 3-bit address. When the IOB voltage is HIGH, the PDEN function is activated. This is active LOW to enable the data transceiver for port actions, performing the same action for I/O as DEN carries out for memory.

The inputs from the 8088 consist only of the S0–S2 status bits, and the remaining inputs consist of the clock pulses and three control inputs which are delivered from other chips. AEN is the address enable input, active LOW, which enables the command outputs of the 82C88 chip. The AEN signal is derived from the 8237A DMA controller, and is used when the bus controller must yield control of the buses for DMA purposes. When AEN is HIGH, the command output drivers are floated, but when I/O is being used the AEN signal has no effect.

The CEN input is active HIGH, and enables both the command outputs and the DEN and PDEN control outputs. With CEN LOW the command lines are inactive (not floating), and the input was intended to allow memory partitioning and to eliminate addressing conflicts. The provision has not been used to any extent. The IOB pin action has already been covered.

Bus arbitration

When the local bus can be controlled by more than one master, as can occur when a co-processor is used, control signals must be available to ensure that only one bus master is active at a given time. On the 8088 type of processor, this is achieved by using the HOLD, HLDA (minimum mode) or RQ/GT (maximum mode) along with the LOCK system.

In minimum mode, the HOLD/HLDA signals are used. HOLD can be taken active HIGH at any point in a bus cycle, but must be synchronized. The HOLD set-up time from the start of HOLD to the next clock pulse must be at least 20 ns for the faster 8088 chips, at least 35 ns for the slower varieties. If this set-up time cannot be guaranteed, external synchronization will have to be used; the use of a suitable co-processor usually ensures that the set-up time will be met. At least one clock state will then elapse before the HLDA signal output is obtained in the middle of a T4 or Ti state. By the time HLDA is valid, the buses will have been floated, and are available for use by a co-processor. When HOLD is reduced, there will again be a delay of 1 or 2 clock states before HLDA is also lowered, following which the buses will be released by the co-processor.

In maximum mode, bus arbitration is carried out using the request/grant signals, which exist in two pairs. The RQ/GT0 pair has priority over the RQ/GT1 pair if any conflict of requests should arise. Unlike the HOLD/HLDA pins, the RQ/GT pins are bidirectional, used alternately for request and grant. The timing of the request/grant sequence was illustrated in Section 5.

The LOCK output is used in maximum mode only, and allows the 8088 to retain bus control by a software command. This allows for software instructions to run without being suspended by a co-processor, or having data that is being transferred through the processor become transferred instead to the co-processor. The LOCK is an instruction **prefix**, meaning that a software instruction can be written with the word LOCK ahead of the instruction, adding a byte to the code that will apply the LOCK action during the whole of the instruction. When LOCK has been used, the LOCK pin output will go low in the next clock state after the LOCK instruction has been decoded. This prevents any bus request from being granted until LOCK is released at the end of the last bus cycle of the instruction that was prefixed with LOCK. Any pending bus request will then be serviced.

One final control signal is the TEST pin, active LOW. This is used as a simple alternative to interrupts for causing a delay, making use of the software WAIT command. If WAIT is executed while the TEST pin is HIGH, then the WAIT instruction will be repeated until the TEST pin voltage is LOW. During this time, the buses can be used by a co-processor, or an interrupt can be serviced, and the WAIT will be resumed if the TEST pin is maintained HIGH.

The 80C88A and 80C88AL use 'bus-hold' circuits to eliminate the need for pull-up or pull-down resistors. When CMOS devices are connected to buses, floating lines can cause high currents to flow, and it is more desirable to ensure that lines which are floating are at one logic voltage or the other. The bus-hold circuits of the 80C86A maintain the last-used logic state on any pin that is unconnected or floating. Any drive to such pins must be able to sink or source at least 350 μA at logic voltage levels. Since the bus-hold action is achieved by active circuits, the power required is considerably lower than when pull-up or pull-down resistors are used.

Memory organization

Circuits for the 8088 can use memory constructed as one bank of 512K of RAM and 8-bits wide, allowing the same circuit techniques as are used for 8-bit microprocessors. RAM can be used above 512K, but the highest memory addresses must be used by ROM (or RAM to which ROM data is copied) because the chip will start reading from address FFFF0H at switch-on or after any RESET. In addition, the addresses 0H to 3FFH must be reserved for interrupt vectors. Each of these consists of 4 bytes, a 2-byte segment number and a 2-byte offset number which will be combined in the usual way. This makes up to 256 possible interrupt vectors available. Some of these are reserved by Intel, others for the operating system in computers.

8086 bus

The 8086 uses time multiplexing for lines AD0–AD15, with addresses A16–A19 multiplexed with status outputs. As for the 8088, these bus lines can be used directly, retaining the multiplexing, or the signals can be demultiplexed using latches and transceivers to give a 20-bit address bus and a 16-bit data bus as illustrated in Section 12.

One control bus output that is implemented on the 8086 and not on the 8088 is the BHE (pin 34), shared with S7. The BHE signal is available during T1 and is active LOW to enable data to be used on the more significant half of the data bus, D8–D15. In a memory read of 8-bit devices, for example, BHE would be high on the first 8-bit read and low on the second 8-bit read. In T2–T4, the pin output is the S7 status signal.

The 80C86A and 80C86AL use 'bus-hold' circuits to eliminate the need for pull-up or pull-down resistors. The details are as for 80C88A and 80C88AL.

Memory organization

The memory is constructed conventionally from 8-bit units, and is organized as a low bank (D0–D7) and a high bank (D8–D15), each consisting of 256K of byte-size memory addressed in parallel by the 20 address lines. Addresses A1–A19 convey even address numbers and are used when the D0–D7 data lines are active. When A0 is HIGH, the address number is odd and the D8–D15 lines will be used. The combination of the BHE and A0 signals allows selection of reading from or writing to odd or even byte locations, or to both in the form of a word read or write. Instructions are fetched as words and bytes are selected internally as needed (if an instruction consists of a single byte, for example).

When data consists of words, the bus action will require one memory cycle if the word starts on an even address number, and two bus cycles if the word starts on an odd address number. Software should therefore be organized to store data in word units located on an even address. The use of an even address number is particularly important for stack operations, because using the stack with odd address numbers can slow down interrupt processing or multitasking actions.

The use of memory addresses FFFF0H for ROM, and of 0H to 3FFH for interrupt vectors follows the same pattern as for the 8088.

80188 bus

The 80188 bus uses time-multiplexed lines for AD0–AD7, address-only lines for A8–A15, and multiplexed address/status lines for A16–A19. The pattern of address and data bus lines therefore follows those of the 8088 but the pattern of control bus lines is different, reflecting the built-in support chips that are a feature of the 80188.

For control of the local bus (between the 80188 and any buffers, latches or transceivers) the 80188 uses the ALE, RD and WR control bus signals whose actions have already been described. There is, however, no M/IO output, so that switching between memory and port use requires one of three possible methods:

- latching the S2 signal;
- making I/O use memory addresses;
- using the built-in chip-select signals.

Transceivers on the data lines can be controlled by the usual DT/R and DEN signals as used in minimum mode of the 8088. The bus arbitration system makes use of the HOLD/HLDA system as used by the 8088 in minimum mode. Since there is only one HOLD input, external chips must be used to determine priority if there is a possibility of more than one co-processor requesting the buses at any given time. In a bus HOLD, the address and data buses are floated, along with DEN, RD, WR, S0–S2, LOCK, S7 and DT/R.

The HOLD request has a high priority, and the bus will be handed over as soon as possible. This implies a time (the **latency** time) which can be as little as one clock state if the HOLD is issued before the start of an instruction, but which can be up to four bus cycles (16 clock states). Long latency times are caused when the HOLD is issued at the start of a DMA transfer of a word from one odd-numbered address to another odd-numbered address, or by the existence of a LOCK action at the time when HOLD is issued.

The other control signals of the 80188/80186 series have been dealt with separately in Section 4. See also the variations applicable to the CMOS versions of

80188 and 80186 under the heading enhanced operation below.

80186 bus

The 80186 bus closely follows the pattern of the 80188 bus, and the differences follow the same pattern as those between the 8088 and 8086. The 80186 uses time multiplexing for lines AD0–AD15, with addresses A16–A19 multiplexed with status outputs. As for the 80188, these bus lines can be used directly, retaining the multiplexing, or the signals can be demultiplexed using latches and transceivers.

The BHE signal, not applicable to the 80188, is available during T1 and is active LOW to enable data to be used on the more significant half of the data bus, D8–D15. In a memory read of 8-bit devices, for example, BHE would be high on the first 8-bit read and low on the second 8-bit read. In T2–T4, the pin output is the S7 status signal.

Enhanced operation – CMOS versions

The 80C188 and 80C186 are pin-compatible with the 80188 and 80186 versions, but have enhanced features such as provision for DRAM refresh and for power-saving operation. These processors can operate either in compatible mode, in which they are almost totally compatible with the 80188/80186; or in enhanced mode which offers the additional features. The only incompatibility in compatible mode is that the 8087 co-processor **cannot** be used, so that the 80C188/80C186 chips are not intended for computing applications in which floating-point processing would require co-processor support. The enhanced mode actions are not available in compatible mode.

Enhanced mode is made available by hardware – the RESET output has to be connected to the TEST/BUSY input. In this state, pin 64 combines the actions of S7 and RFSH. The voltage on this pin will go LOW to indicate a memory refresh cycle. Refresh is controlled by registers within the processor which include a counter to set the time between refresh cycles. After each of these time intervals a memory read is carried out in order to refresh memory. The memory read request will put HLDA LOW if a HOLD is in use, and this change of HLDA can be used to signal to an external device that is using the buses that a memory refresh is due. The HOLD still has priority, and it is for the software to decide when the refresh will be carried out.

The registers that control DRAM refresh cannot be written in compatible mode, and always read out as zeros in compatible mode. In enhanced mode, these registers can be written, using their offset addresses, and read. The address that is generated during a refresh cycle is decided by the contents of the MDRAM register, offset E0H, plus the contents of a 9-bit counter, the refresh address counter. The highest 7 bits in the register should correspond to the chip-select address that has to be used, and the lower nine bits are reserved. The resulting address is as shown in Figure 6.1, with 6 upper bits from the register, 3 zero bits and 1 unity bit, plus 9 bits from the counter. The time interval between refresh cycles is controlled by the CDRAM register, offset E2H, and the other register used is the enable register (EDRAM), offset E4H. The highest bit of this latter

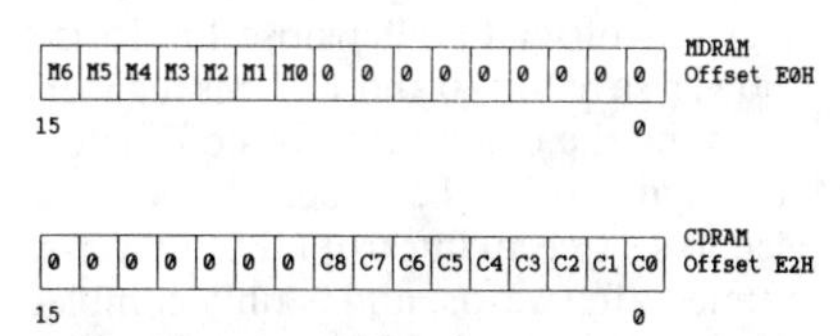

NOTE: Bits 0 - 8 are C0 - C8 comprising clock divisor register and holding the number of CLKOUT cycles between each refresh request.
Bits 9 - 15 are reserved and always read as zero.

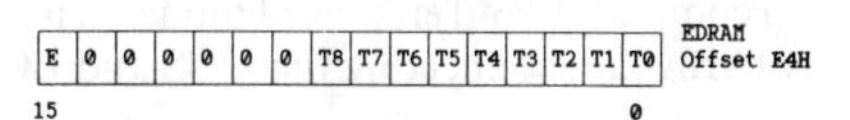

NOTE: Bits 0 - 8 are refresh clock counter outputs, read only.
Bits 9 - 14 are reserved, read as zero
Bit 15 enables Refresh control unit, is reset to 0 on RESET.

Final Generated Address

M6	M5	M4	M3	M2	M1	M0	0	0	0	CA8	CA7	CA6	CA5	CA4	CA3	CA2	CA1	CA0	1

A19 ... A0

M6-M0 are bits obtained from MDRAM register
CA8-CA0 are bits obtained from the refresh address counter

Figure 6.1 *Using the refresh registers of the 80C188, 80C186, a feature of enhanced operation for these chips that is not present on the 8088.*

register is the E bit, and setting this bit to 1 enables the refresh control unit (RCU).

Setting the E bit starts a refresh cycle, and in T3 of the bus cycle in which the E-bit has been set, the clock counter bits will be loaded from the CDRAM register into the lowest 9 bits of the EDRAM register. Each CLKOUT pulse will then decrement this latter register. When the counter value reaches 1, a refresh will be requested and the count is restarted. This will continue, keeping memory refreshed at the stipulated intervals, until the E-bit is cleared (which does not reset the address counter).

The power-save action is the other feature of enhanced mode. This operates by slowing down the internal clock, allowing the processor to work at a lower speed and a correspondingly lower power level. The internal clock speed must **not** be allowed to fall below 0.5 MHz. Because the reduced clock rate affects all chip actions, the DRAM refresh registers must be reprogrammed each time this feature is switched on or off, since memory refresh cannot be slowed down. The use of power-save is more applicable to systems that use CMOS static memory rather than DRAM.

Power-save is controlled by the contents of the PDCON register, offset FOH, whose bit 15 is the enable for power-save. Bits 0 and 1 of this register are used to determine the clock rate division factor, as indicated in Figure 6.2. The processor returns to normal speed to service an interrupt, but power-save speed is resumed afterwards, and is only completely abandoned when the enable bit is reset.

One other feature of the CMOS variants of the 80C188 and 80C186 type is their test mode, applicable in both compatible and enhanced modes. This is selected by hardware, connecting the UCS and LCS

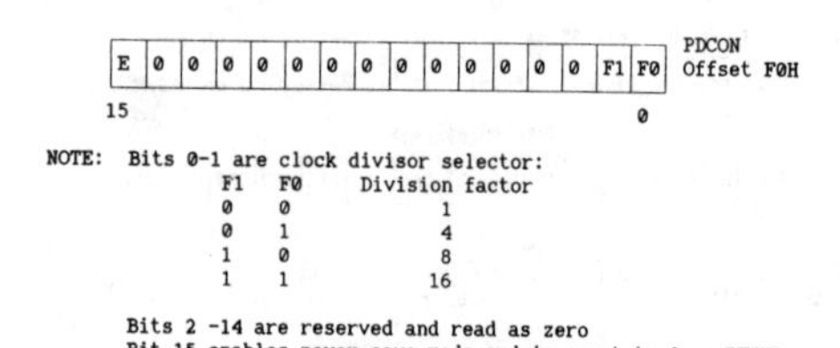

Figure 6.2 *Controlling the power-save feature using bits in the PDCOM register.*

pins LOW during a RESET. When this is done, all pins are floated until the next RESET (with UCS and LCS released).

80286 bus

The local bus of the 80286 consists of the address, data, status and control signals of the processor, and the system bus consists of signals that have been buffered or latched. The distinction is necessary because the signals on the system bus are likely to differ (some status signals will be absent, some control signals different) and may have different timings. The local bus is normally taken to consist of the 24-line address bus, the 16-line data bus and eight of the status and control signals.

Address and data

The 80286 bus set uses a simple unidirectional bus, active HIGH, for lines A0–A23. The data bus is a bidirectional bus of 16-bits width, and both address and data buses float during a hold acknowledgement, as do many of the control signals. The normal use of the 80286 makes use of a set of chips to establish a system bus. This set of chips is composed of the 82C284 clock generator, the 82288 bus controller and (optionally) the 82289 bus arbiter, along with transceivers and latches. Of these chips, the 82C284 generates the system clock pulses and also synchronizes READY and RESET signals. The 82288 decodes the status signals from the 80286 into a full set of bus control signals. The 82289 chip is used to generate bus arbitration signals for the IEEE 796 standard multibus. See Sections 7 and 12 for the use of chips to create the AT computer bus.

A bus cycle consists of at least two processor cycles, equivalent to four system clock cycles. There are two phases to each processor clock cycle, corresponding to the negative transitions of the system clock voltage. Four bus states are available, the send state Ts, command state Tc, idle state Ti and hold state Th. The Th state is used while the 80286 has given the use of its local buses to a co-processor. Each bus state endures for one processor clock, two system clock cycles. The state table for the 80286 bus states has already been dealt with in Section 5.

Bus control

The control signals from the 80286 are intended to be used by the 80288 bus controller – there is nothing corresponding to minimum mode of the 8088/8086 family. The 80288 provides the ALE, DT/R and DEN signals whose functions have already been discussed, and these signals allow control of all address latches, data transceivers, write and read enable for memory or port uses.

Two signals are provided on the local bus for altering the system timing, providing command extension and command delay. The READY input signal (active LOW) allows for repeating T2 states until external chips are ready, and this is the action that corresponds with the insertion of wait states in the 8088/8086 series. By using external controllers for READY, each bus operation can be extended for as long as is necessary – this also implies that no bus operation need be extended for longer than is needed for the external chips to respond.

Command delay is provided by the CMDLY signal into the 82288 chip. Following the Ts state, each trailing edge of the system clock is used to sample the CMDLY input and if this input is HIGH, the command signal (RD or WR) is not activated. The use of CMDLY can reduce the read or write time that is available, so that the use of CMDLY, if it is used at all, should be confined to a restricted set of bus operations, using an address decoder to enable the CMDLY input.

The bus status signals exist only in the Ts state of a bus cycle. A Tc state always follows the Ts state, and the Tc states will be repeated, forming a wait, until the READY signal goes LOW. The READY signal will be obtained by way of the 82C284, which will ensure that the READY signal is synchronized to the system clock. The 82C284 provides both ARDY and SRDY inputs for this purpose, see Section 7.

Priority on the local bus

The order of priority on the local bus is important, because the bus is used by several chips, and can be subjected to HOLD requests from chips external to the bus. If more than one demand is made simultaneously, the order of priority is as follows, starting with the highest priority:

1 Any use of LOCK in a software instruction, or the use of an action which will enforce a LOCK, such as an access to segment descriptor, interrupt acknowledge, or the XCHG with memory.
2 The second of a pair of single-byte bus actions that are required because a word has started at an odd address number.
3 The second or third cycle of a data transfer to or from a co-processor.
4 A HOLD input to request local bus use.
5 Co-processor data operand transfer using PEREQ input.
6 Data transfer by the execution unit as part of an instruction.
7 An instruction pre-fetch request from the bus unit.

Bus arbitration

When the local bus can be controlled by more than one master, as can occur when a co-processor is used, control signals must be available to ensure that only one bus master is active at a given time. On the 80286 processor, this is achieved by using the HOLD and HLDA system as has been described previously, along with specialized signals for use with the 80287 co-processor. This implies that when the 8087 is used, the HOLD/HLDA system is still available to allow another unit to take charge of the buses, subject to the priority order noted above. The use of the numerical co-processors is dealt with at the end of this section. The 80286 supports the use of the LOCK output, following the same pattern as has been described earlier.

Memory organization

Circuits for the 80286 can use physical-present memory up to a maximum of 16 Mb in protected mode, up to 1 Mb in real mode. Software can be used to make some of the memory beyond the 1 Mb limit available to a processor in real mode, but this does not make the additional memory (called **extended memory**) available to all applications programs, only to those which are written to take advantage of it. The usual organization of memory into bytes is dealt with in the same way as on the 8086, by allowing byte transfer on the other half of the data bus. The same comments on even or odd alignment of words apply to the 80286.

After a RESET, the 80286 will be placed into real mode, and when the CS register is in use, as it will always be for a code fetch, the address lines A20–A23 will be high, so that the initial memory call will be to the address written as F000: FFF0, corresponding to the 20-bit address FFFF0. This provides 64K of space for the first code that the processor will read after switching on.

If the processor is to be used in protected mode (something that is not done on computers using the 80286), the GDT and IDT registers (see Section 2) must be loaded with valid numbers. The LMSW instruction will be used to switch to protected operation, and then the processor must be forced to execute a JMP between segments in order to clear the queue of instructions that will have been accumulated while working in real mode. The JMP should be carried out with reference to the first task state segment (TSS) that is to be used.

Returning from protected mode to real mode was not adequately considered in the design of the 80286. In practice it has been done by changing the sizes of both GDT and LDT to zero and then task-switching. The error which is generated by a lack of descriptor will then cause a switch to real mode. In the PC-AT type of computer a call to INT 15H allows for a real-mode command to move data to or from protected mode. For details of these methods, see *Advanced OS/2 Programming* from MicroSoft Press.

The 80386DX bus

Address and data

The 80386 bus set uses a simple unidirectional bus for lines A2–A31, with the lower bits of any address provided from the bit address pins, active LOW, BE0–BE3. This allows address numbers to be incremented by 4 bytes (double-word) so as to read or write 32-bit instructions or data, with the BE inputs held high. It also allows considerable flexibility in using byte or word data, and a byte can be located in any part of memory by making use of the BE outputs. The total physical address space is 4 Gb,

BE3#	BE2#	BE1#	BE0#	D24-D31	D16-D23	D8-D15	D0-D7	Auto dup'n?
1	1	1	0	U	U	U	A	NO
1	1	0	1	U	U	B	U	NO
1	0	1	1	U	C	U	U	YES
0	1	1	1	D	U	U	U	YES
1	1	0	0	U	U	B	A	NO
1	0	0	1	U	C	B	U	NO
0	0	1	1	D	C	D	C	YES
1	0	0	0	U	C	B	A	NO
0	0	0	1	D	C	B	U	NO
0	0	0	0	D	C	B	A	NO

NOTES: D - write data D24-D31
C - write data D16-D23
B - write data D8-D15
A - write data D0-D7
U - Undefined

Figure 6.3 *The byte enable outputs of the 80386DX allow any data alignment to be used.*

corresponding to addresses 00000000H to FFFFFFFFH, along with 64K of port space 0000H to FFFFH.

Each byte enable output allows a different set of bits on the data bus to be used for a single byte transfer, so that BE0 controls D0–D7, BE1 controls D8–D15, BE2 controls D16–D23 and BE3 controls D25–D31. The use of 16-bit buses is made easier by automatic duplication, in which writing an operand that occupies **only** the upper word of the data bus results in that word being duplicated on the lower word, D0–D15. The pattern of the byte enable outputs indicates the type of transfer that is being made, as Figure 6.3 indicates. The conventional A0 and A1 address signals can be generated if required from the BE0–BE3 signals by gating, as Figure 6.4 shows.

Dynamic data bus sizing

The 80386DX uses a 32-bit set of data pins, but these can be connected to either a 32-bit bus, to a 16-bit bus, or to both sizes of bus in one system. The data pins may also be connected to either 32-bit or 16-bit ports. This is made possible by determining the required bus width during each bus cycle, using the BS16 input signal (80386DX only). In each bus cycle, the device that is using the data bus can, by using address decoding or by way of controlling the BS16 line directly, put the BS16 voltage LOW for 16-bit bus operations or HIGH for 32-bit operations.

When BS16 is LOW, any operands that consist of more than 16-bit are automatically dealt with by

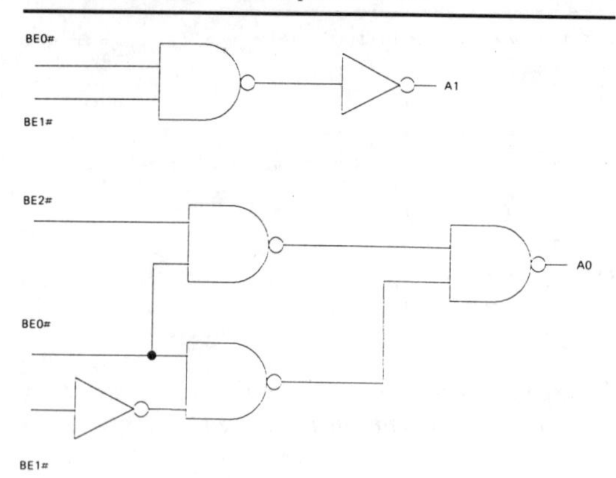

Figure 6.4 *Generating conventional A0 and A1 signals from the byte enable signals.*

more than one transfer, and this applies also to misaligned transfers (any bus address in which a byte enable signal is LOW). In this state, all transfers are made over D0–D15 only. When BS16 is to be selected for one cycle only, the voltage on BS16 must go low when either BE2 or BE3 (or both BE2 and BE3) is LOW. If only single word transfers are being made over the D0–D15 lines, it does not matter whether BS16 is used or not – the use of BS16 applies when more than 16 bits are being read and used and must be read in more than one action. For transfers involving the lower word of a 32-bit set, the BE2/BE3 signals would not be used.

The effect of switching BS16 on (LOW) for a 16-bit transfer depends on whether the cycle is read or write, and on whether the upper half of a set of bytes is to be read, or both upper and lower half. The summarized action is as follows:

Read, upper half only: The 80386DX reads data on the lower 16-bits of the data bus, ignoring the upper 16-bits. Data that would have been read from D16–D31 is read from D0–D15 instead, using these lines for the upper half of a double-word.

Write, upper half only: The use of BS16 has no effect, because when only BE2 AND/OR BE3 is used, the signals on D16–D31 are automatically duplicated on to D15–D0.

Read, upper and lower: Bytes 0 and 1 are read along D0–D15 on the first cycle, and bytes 2 and 3 along D0–D15 on the second cycle. D16–D31 voltages are ignored.

Write, upper and lower: All bytes are placed on the data bus on the first write cycle, so that bytes 0 and 1 are on lines D0–D15. On the second cycle, bytes 2 and 3 are automatically duplicated from D31–D16 to D15–D0, providing the correct 16-bit write action.

Word alignment

When the 80386DX is used along with 32-bit wide memory (at the time of writing this is not common) then each double-word (Dword) of data starts with its byte 0 at an address number that is divisible by 4. The address bits A2–A31 are used to locate Dword positions, and the BE bits, if necessary, to select bytes or words within a Dword. The BS16 voltage is held HIGH to enforce 32-bit bus use.

In the more common situation where the 80386DX is used with 16-bit memory (common because software for 80386 computers is mainly software written earlier for 8088/8086 machines), each 16-bit word will start with its byte 0 located at an address number that is divisible by 2. For such systems, the BS16 input can be taken permanently LOW unless 32-bit ports are being used, which is unlikely. For most 16-bit memory and port systems as used on 80286 and 8086 machines, byte enable signals will have to be generated so that one bank of memory bytes can be activated to be used as low order bytes and another bank to be used as high order bytes. The BHE and BLE signals that are needed for this purpose can be generated from the BE0–BE3 signals from the 80386DX processor. The A1 signal is required as well, and gating for that signal has already been illustrated, along with gating for A0 which is also the BLE signal. Figure 6.5 shows gating that can be used to derive the BHE signal.

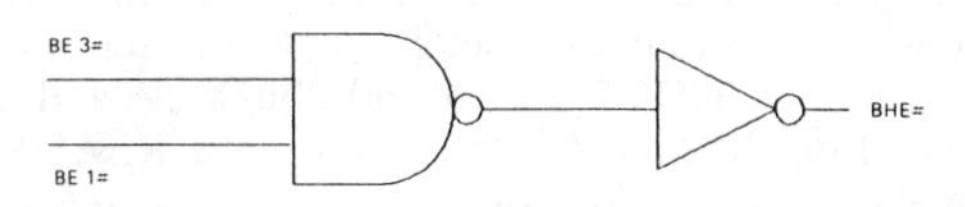

Figure 6.5 *Gating the BE3*# and BE1# signals to obtain BHE#.

One problem that can arise is an anomalous byte order which can occur when the address in memory is the last byte of a Dword, an address whose A0 and A1 bits are both 1. On a 32-bit bus, a word located at such an address will be transferred in 2 bytes, **high byte first**. This is the opposite of the byte order used in all of the earlier processors, so that such transfers should be avoided. There are no anomalies concerning byte or Dword transfers over 32-bit buses, or with any transfers over 16-bit buses.

Control and definition signals

The bus cycle definition signals are W/R, D/C, M/IO and LOCK, each of which acts to define the type of bus cycle that is being performed. W/R is HIGH for write, LOW for read, and D/C is HIGH for data and LOW for control. The M/IO signal is HIGH for a memory access and LOW for a port (I/O) access. LOCK is put LOW to ensure that the local bus is not made available to any other chip. The correspondence between these signals and the bus state was shown in Figure 4.30 for the first three of these signals, each of which is valid at the same time as ADS is driven LOW. Note that there is one combination which does not occur in a normal bus cycle, but which can appear in an idle state when ADS is **not** low.

The LOCK signal goes LOW at the time when the first locked bus cycle starts, but this is not necessarily coincident with ADS, because LOCK is a response to the software LOCK instruction which, if pipelining is being used, could be decoded and acted on before ADS is active. LOCK is always ended when the READY signal goes LOW on the last locked bus cycle.

The bus control signals consist of ADS, READY, NA and BS16, all active LOW. The address status ADS signal can be used for latching address numbers, and is needed in particular when pipelining is used and each address changes midway through a cycle. The ADS output goes LOW in the T1 and T2P states – see Section 5 for details of timing and state sequences. READY is used to end a bus cycle, and is ignored on the first state of a cycle. The READY signal must be provided from external chips, such as the 82C284 clock generator, so that the synchronization is applied to meet the set-up time requirements.

The use of the NA and BS16 signals has already been covered in detail.

Bus arbitration

The 80386DX is designed to be used along with co-processors, and a complete set of special interfacing signals can be used for the 80387 or 80287 co-processors. In addition the HOLD and HLDA system which has been used on all of the other processors in the family is also available, and the action need not be described again. The interfacing with the floating-point co-processor will be described later in this section.

80386SX bus

The 80386SX differs from the 80386DX (or 386) in using a 16-bit data bus and a 24-bit address bus rather than the 32-bit buses of the 80386DX, though the internal architecture uses the 32-bit structures of the 80386DX and the chip is software-compatible.

The address lines consist of A1–A23 for access to even-numbered memory addresses, plus two byte enable signals for selecting odd or even address positions. This allows 16 Mb of physical memory to be used, along with 64K of port addressing. The byte enable bits BHE and BLE, both active LOW, determine the type of transfer (Figure 6.6).

In general, the 80386SX is identical in bus action to the 80386DX, apart from the use of restricted bus width. All data transfers are either word or byte sized, and there is no BS16 pin needed or provided. The LOCK action is restricted to the instructions shown in Figure 6.7, and is not supported on the repeated string actions as it is on the 80286. Illegal use of LOCK will cause a type 6 exception to be generated.

Byte Enable	Data Bus Signals	
BLE#	D7 - D0	(least significant byte)
BHE#	D15 - D8	(most significant byte)

Figure 6.6 *Using the byte enable signals on the 80386SX.*

Opcode	Operands - To,From
BIT test and SET/RESET or COMPLEMENT	Memory reference, register/immediate
XCHG	Memory reference, register
XCHG	Register, memory reference
ADD,OR,ADC,SBB,AND,SUB,XOR	Memory reference, register/immediate
NOT,NEG,INC,DEC	Memory reference

Figure 6.7 *The use of LOCK on the 80386SX is restricted to this set of instructions.*

Floating-point co-processors

Section 7 deals with the support chips for the Intel chips, but the floating-point co-processors fall into a class of their own because their presence is optional, and their relationship with the main processor is so close that they can be considered as a local bus extension of the main processor. The co-processors that are currently available are the 8087, used with the 8088, 80188, 8086 and 80186 processors, the 80287 which can be used with the 80286 and the 80386, the 80387 which is for the 80386 chip, and the most recent addition, the 80387SX which is specifically intended for use with the 80386SX processor. In this section we are concerned primarily with the interfacing of these chips to the main system, rather than with their instruction sets.

Each of these chips adds floating-point routines for simple arithmetic, plus trigonometric, exponential and logarithmic functions, using up to 80-bit number representations and 18-bit BCD. The chips are compatible with the IEEE floating-point standard 754. The speed-up of floating-point operations is achieved in two ways; by the use of hardware that is not present in the CPU along with very efficient coding, and by the fact that the co-processor can work on floating-point operations while the main processor executes other instructions. Note that the co-processor takes charge of the buses while transferring data, so that the main CPU is idling in this time – the CPU and the co-processor cannot load and save in parallel. None of the chips from 8088 to 80386 provides for floating-point arithmetic, and this is performed either by the co-processor or by software. The i486 (80486) includes floating-point provision and needs no floating-point co-processor.

The parallel operation of CPU and co-processor can cause conflicts of the timing of instructions. If, for example, the co-processor has a single instruction to place a floating-point number into memory, and this is followed by an instruction to the CPU which reads from the same memory location this will lead to the wrong value being read by the CPU. The reason is that the timing of CPU and co-processor provides that if a CPU instruction immediately follows a co-processor instruction, the two will be performed in parallel, and in this example, if the co-processor takes more time to execute its instruction then the second instruction (to the CPU) will be performed first. The solution is to insert a wait instruction for the 8087 processor (FWAIT) between the adjacent memory-load instructions. Assembler programs for the older chips will insert these wait instructions automatically.

The 8087

The 8087 is a 5 MHz chip, and faster versions are designated as 8087–2 (8 MHz) and 8087–1 (10 MHz). The 8087–1 is also suitable for use with systems that use the 80188 or 80186 at 8 MHz. These chips are extremely expensive, about 25 per cent of the price of a complete 8088-based computer, and are used only if the use of the computer demands a lot of floating-point arithmetic work – examples are the extensive use of spreadsheets and CAD. For many purposes, the user of any 8088/8086 machine should contemplate the alternative of upgrading to an 80286 or 80386 machine, which even with no co-processor will handle floating-point operations much faster than the 8088 or 8086 machines.

The pin-out is shown in Figure 6.8, using a conventional 40-pin DIL package, and the layout of the pins follows very closely that of the 8086. Address and data pins are multiplexed from AD0 to AD15, and the remaining four address pins are multiplexed with status bits as on the 8086. The other pins provide actions that should already be familiar from the 8086 architecture.

The block diagram for the 8087 is illustrated in Figure 6.9, with the division into control unit and numeric execution unit. The control unit governs the transfer of data between the co-processor and the

Pin configuration - the # sign indicates active LOW.

1. GND	40. V_{cc}
2. A14	39. A15
3. A13	38. A16/S3
4. A12	37. A17/S4
5. A11	36. A18/S5
6. A10	35. A19/S6
7. A9	34. BHE#/S7
8. A8	33. RQ#/GT1#
9. AD7	32. INT
10. AD6	31. RQ#/GT0#
11. AD5	30. NC
12. AD4	29. NC
13. AD3	28. S2#
14. AD2	27. S1#
15. AD1	26. S0#
16. AD0	25. QS0
17. NC	24. QS1
18. NC	23. BUSY
19. CLK	22. READY
20. GND	21. RESET

Figure 6.8 *Pin allocation for the 8087 floating-point co-processor, which uses a conventional 40-pin DIL package.*

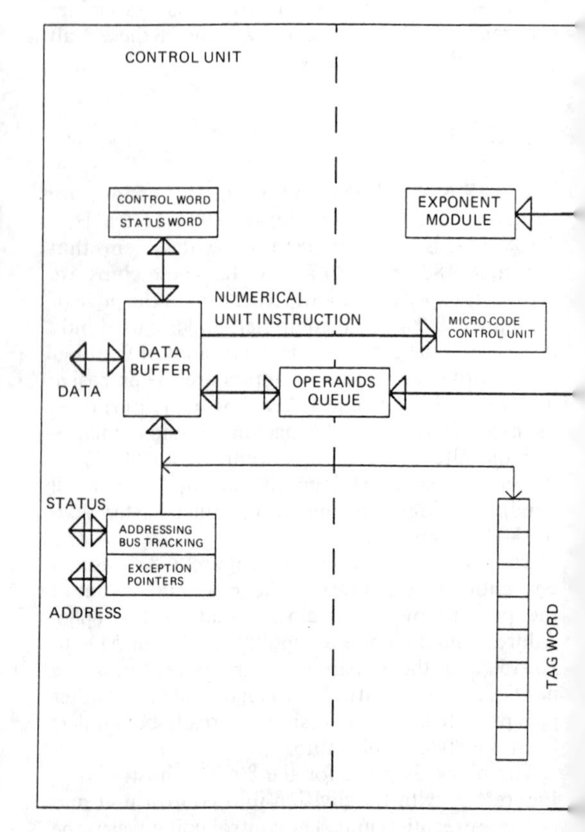

Figure 6.9 *The block diagram for the 8087.*

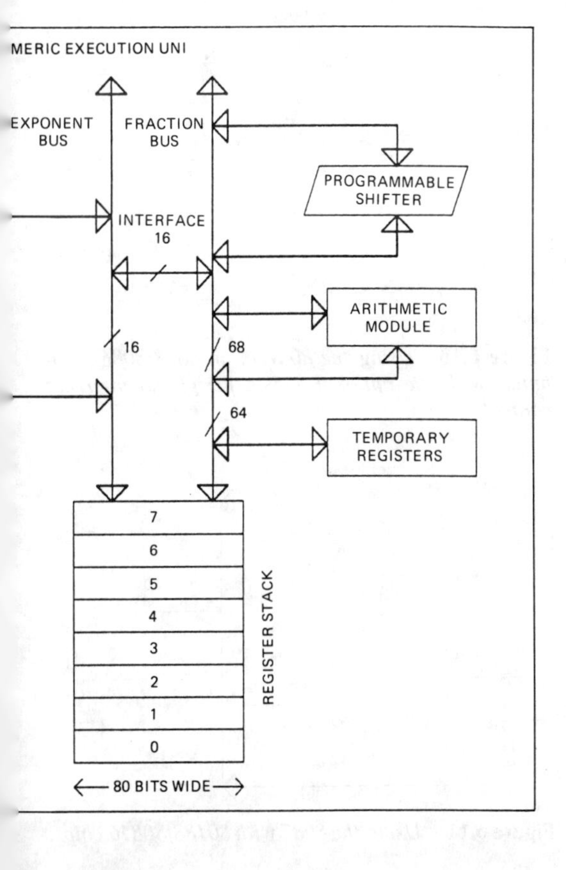
MERIC EXECUTION UNI
EXPONENT BUS
FRACTION BUS
PROGRAMMABLE SHIFTER
INTERFACE 16
ARITHMETIC MODULE
16
68
64
TEMPORARY REGISTERS
7
6
5
4
3
2
1
0
REGISTER STACK
80 BITS WIDE

main processor, and the feeding of data and instructions to the numeric execution unit. This latter unit features two buses, one of 16-bits for the exponent of a floating point number, and the other of 68-bits for the mantissa (the fraction). A floating point number is represented by a fraction which is multiplied by a power of 2, with the power value represented by the exponent value. The chip contains 8 registers, each of 80 bits, which can be individually read and written from the main processor.

Connections

Figure 6.10 shows the 8087 connected to an 8088 system, and Figure 6.11 shows the connection with

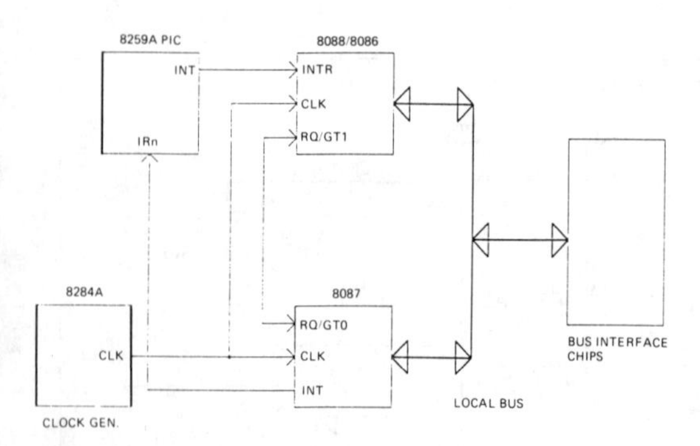

Figure 6.10 *Using the 8087 in an 8088/8086 computer, with the optional 8259A peripheral interrupt controller chip.*

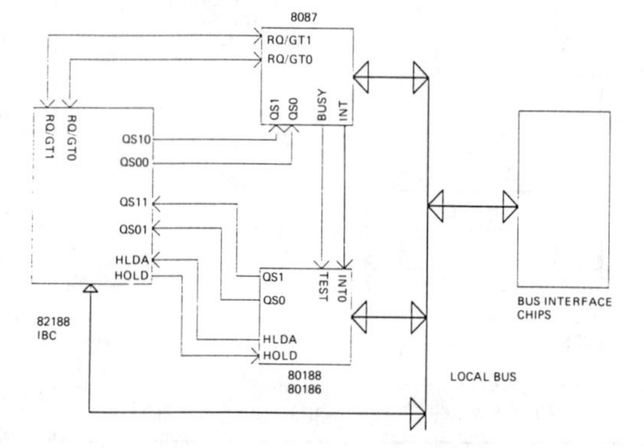

Figure 6.11 *Using the 8087 with 80188/80816 chips.*

an 8086 system. The connections for 80188/80186 follow along similar lines, with the supporting chips of the 8088/8086 set replaced by the signals from the 80188/80186 chips directly.

Registers

The registers of the 8087 are illustrated in Figure 6.12. The data field registers consist of eight 80-bit registers, each of which is associated with a 2-bit 'tag-field' in a separate tag register. The other seven registers, each of 16 bits, consist of control register, status register, tag word register, along with two registers each for instruction pointer and data pointer.

Taking the 16-bit registers first, the control register contains the control word for the 8087, whose bit breakdown is shown in Figure 6.13. The lower byte of this word is concerned with interrupts and exceptions, and each bit can be set to allow an interrupt or

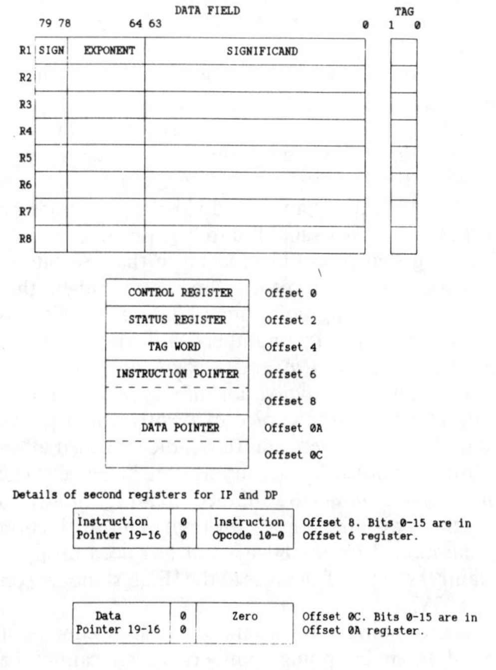

Figure 6.12 *Registers of the 8087 co-processor.*

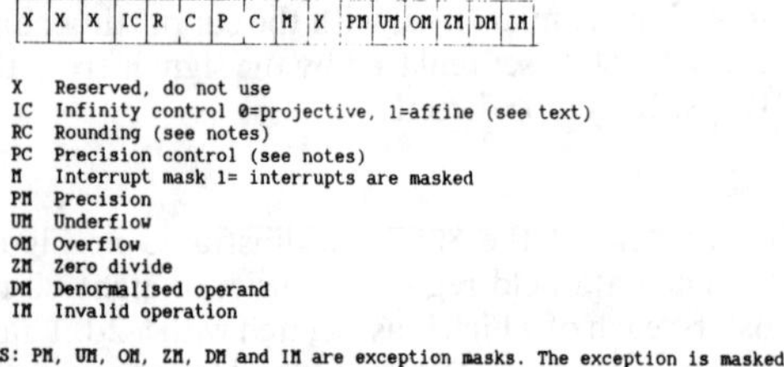

Figure 6.13 *The bit diagram for the control word of the 8087.*

exception to be masked (disabled) – when the low byte consists of zeros, all interrupts and exceptions can be executed. Bits 0 to 5 allow for the masking of the six specific interrupts and exceptions that are used specifically by the 8087, and bit 6 is reserved. Bit 7 is a general mask bit, and by setting this bit, all interrupts and exceptions are masked.

The high order byte controls precision and rounding. The three most significant bits are reserved, and bit 12 is the 'infinity control'. An infinite result is produced when a quantity is divided by zero, or by a number which is so small as to be equivalent to zero (meaning that it would need more than 80 bits to produce a non-zero result). Since all numbers that are used in floating-point binary have a sign bit, it is possible to have the numbers $+0$ and -0, and therefore to have $+\infty$ and $-\infty$. Any additions or subtractions that involve an infinity will be differently affected if the choice of positive or negative infinity is maintained, and this choice is called **affine** infinity, signalled by the infinity bit being set. The alternative is **projective** infinity, treating infinity as unsigned, with the infinity bit reset. This distinction is maintained on the 80287, but has been dropped (mainly because of changes to the IEEE standard) on the 80387.

The rounding control bits permit four choices of rounding or chopping numbers which cannot be expressed exactly. The choices are:

00 round to nearest or to an even number if midway
01 round down towards $-\infty$
10 round up towards $+\infty$
11 truncate (chop) number

and the precision control bits 8 and 9 offer the choice of 24-bit, 53-bit or 84-bit precision for the mantissa. The lower precision options are provided only for compatibility with other forms of floating-point computation, and the usual option is to have both bits set to provide full 64-bit precision.

The status word register is illustrated in Figure 6.14. Bits 0 to 5 will be set to indicate that an exception has occured during an operation, and bit 6 is reserved. Bit 7 is the interrupt request bit which will be set if any exception bit is set and not masked; it is otherwise maintained reset. Bits 11–13 are used to point to the register that is the top of the stack, referring to the 80-bit data registers which are addressed on a first-in-last-out basis like a software stack in RAM. Bit 15 is the busy bit which will be set when the unit is processing an instruction or has an interrupt request pending. If the numeric unit is idle, this bit is reset.

Bits 8–10 and 14 form the numeric condition flags, labelled as C0–C3, which are used in much the same way as the flags of a CPU to reflect the results of the most recent action so that testing these flags will allow the co-processor to jump to different parts of its code instructions. The full table of flag use is shown in Figure 6.15. The compare tests show the result of comparing a value, the source, with the value stored in the register wich is currently top of the stack. The remainder tests show whether a division

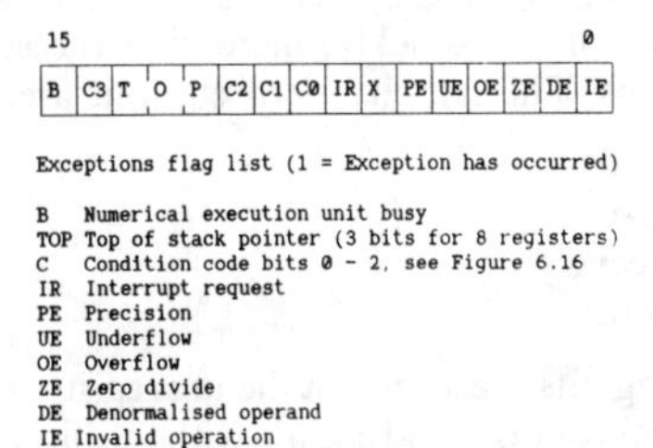

Figure 6.14 *The status word register for the 8087.*

Instruction type	C3	C2	C1	C0	Interpretation
Compare, test	0	0	X	0	Top>source or zero
	0	0	X	1	Top<source or zero
	1	0	X	0	Top=source or zero
	1	1	X	1	Top not comparable
Remainder	Q1	0	Q0	Q2	Total reduction, 3 low-bit Q't.
	U	1	U	U	Incomplete reduction
Examine	0	0	0	0	Valid, positive un-normalised
	0	0	0	1	Invalid, positive, exponent=0
	0	0	1	0	Valid, negative un-normalised
	0	0	1	1	Invalid, negative, exponent=0
	0	1	0	0	Valid, positive, normalised
	0	1	0	1	Infinity, positive
	0	1	1	0	Valid, negative normalised
	0	1	1	1	Infinity, negative
	1	0	0	0	Zero, positive
	1	0	0	1	Empty
	1	0	1	0	Zero, negative
	1	0	1	1	Empty
	1	1	0	0	Invalid, positive, exponent=0
	1	1	0	1	Empty
	1	1	1	0	Invalid, negative, exponent=0
	1	1	1	1	Empty

NOTES: Top means top of register stack (using Reverse-Polish logic)
X - value not affected by instruction
U - value is undefined following instruction

Figure 6.15 *The condition code bits of the 8087 and their meaning.*

has been complete (no remainder) or not, and if complete indicates the low bits of the quotient. The examine tests are used to determine the type of number resulting from an operation. In this connection, normalized means that the mantissa is a binary fraction which represents the fractional part of a binary number starting with 1. In simple terms, a normalized value of binary .001 means that the actual mantissa is binary 1.001, and the purpose of normalizing is that since a normalized number **always** starts with a 1 before the binary point, this 1 need not be stored.

The tag word register is used to maintain a 2-bit tag code for each data register. The tag word shows what the content of a register is, so that the 8087 can check for content quickly, increasing the speed of calculations. The tag bit interpretations are:

00 valid
01 zero
10 special
11 empty

and the tag bits are stored in the tag register starting with the bits of tag 0 and ending with the bits of tag 7.

The instruction and data pointer registers are provided to make it easier for a programmer to write special-purpose exception handlers. When the 8087 executes a floating-point instruction, the instruction address, the operand address and the instruction opcode (its 11 lowest-order bits) are all stored in these registers and can be stored from there into memory if needed.

Signals and bus

The signals at the pins of the 8087 closely follow the signals of the 8088/8086, so that shorter descriptions will be used here. The 8087 would not be used along with an 8088/8086 in minimum mode, so that the use of the 8288 bus controller for 8088/8086 or the 82188 bus controller for 80188/80186 is assumed, and some of the interconnections will be made with these chips, as identified by the naming of pin functions.

DC supplies are V_{cc} of +5 V on pin 40 and earth ground on pin 20. The V_{cc} pin should be decoupled with a 10 nF capacitor. The CLK input, pin 19, accepts the 33 per cent duty cycle clock from the usual 8284A clock generator chip to provide the basic 8087 timing. The READY input, active HIGH, is used to acknowledge a transfer of data from memory or port. Systems that use the 8088/8086 will be able to synchronize a READY signal using the 8284A clock generator, and 80188/80186 systems can use the 82188 integrated bus controller for this purpose.

The RESET input on pin 21, active HIGH, causes the co-processor to terminate any current instruction. The signal must be active for at least four cycles of the system clock, and is synchronized internally. The other inputs are QS1 (pin 24) and QS0 (pin 25) which convey the status information from the CPU to the co-processor, interpreted in the same way as the 8088/8086 in MAX mode (Figure 4.4). These inputs allow the 8087 to keep track of the instruction queue in the CPU, and connect to the corresponding pins of the 8088/8086 in maximum mode.

Of the outputs, the INT (pin 32), active HIGH, occurs when an exception occurs during 8087 processing, and is enabled (not masked in the control word register). This interrupt will normally be taken to the 8259A interrupt controller for 8088/8086 systems, and to the INTO pin of 80188/80186 systems. The other output-only pin is BUSY, pin 23, active HIGH. This connected to the TEST pin of the

CPU and is used to indicate that the 8087 is working on a floating-point calculation – note that the TEST pin has no effect until tested *by software*. BUSY remains HIGH if an exception occurs and is not cleared. By tying these pins, the main CPU is forced to execute idle states if an FWAIT instruction is used. There is no automatic way in which the CPU will check whether the 8087 is ready for input, and so any software that will use the 8087 must place the FWAIT instruction preceding the floating-point instruction to ensure that the TEST pin voltage will be tested. This does **not** apply to other CPU-co-processor combinations.

The remaining pins are used for both input and output. The AD0–AD15 pins are multiplexed address and data pins which provide an address for the T1 state and transfer data on the T2, T3, Tw and T4 states. These address/data lines can be directly connected to the 8088 or 8086 local bus lines. When used with the 8088/80188, the 8087 will maintain an address on the A8–A15 lines throughout states T1–T4, but when used with an 8086/80186 these lines will be multiplexed in the same way as the corresponding 8086 lines. This is done by monitoring the BHE/S7 line after any RESET, and this automatic detection also allows the 8087 to match its instruction queue length to that of the CPU. In the times when the CPU has control of this bus, the 8087 will monitor the content of the data lines.

The highest order address lines A16–A19 are timeshared with the S3–S6 status inputs. The lines are used for address bits in the T1 state, and for status in the remaining bus states. When the 8087 is in control of the buses, S3, S4 and S6 are maintained HIGH and S5 is maintained LOW. The S6 voltage is used by the 8087 to distinguish bus action by the 8088/80188 or 8086/80186 from bus use by any other chip – only the main CPU should be able to drive the S6 signal LOW. The BHE/S7, pin 34, is used for detecting bus width and enabling data to the high byte of the 8086 bus. On the 8088/80188, this pin is maintained HIGH in maximum mode, but on the 8086/80186 the pin voltage goes LOW following a RESET. This is used at each reset to determine how the 8087 uses its data bus, and also to determine the length of the instruction queue.

The S0, S1, S2 status pins (26, 27 and 28) are active

S2#	S1#	S0#	Status report
0	0	0	Unused
0	0	1	Unused
0	1	0	Unused
0	1	1	Unused
1	0	0	Unused
1	0	1	Read memory
1	1	0	Write memory
1	1	1	Passive (no bus cycle)

Figure 6.16 *Status pin decoding for the 8087.*

low in the T4 state of the bus and remain active in the following T1 and T2. They return to the HIGH setting in T3, or during a Tw state (forced by READY being HIGH). These status signals will be used by the 8288 bus controller in 8088/8086 systems or the 82188 integrated bus controller in 80188/80186 systems to generate the control signals for memory access, using the coding indicated in Figure 6.16. Any change in these signals during T4 is used to indicate the start of a new bus cycle, and the return to the HIGH state signals the end of a bus cycle. When the CPU controls the buses, the corresponding CPU signals are monitored by the 8087.

The main interactions between the CPU and the 8087 are made by way of the RQ/GT pins, which must be connected to the corresponding pins of the 8088/8086 in maximum mode. The 80188/80186 processors use the HOLD/HLDA bus control system, and this has to be converted to the RQ/GT format using the 82188 integrated bus controller. Systems using the 80188/80186 can make use of three processors, consisting of the 80188/80186, the 8087 and another co-processor such as the 82586 LAN co-processor or the 82730 text co-processor. Such additional co-processors may use either HOLD/HLDA protocols, or RQ/GT, and either type can be connected, using the HOLD/HLDA type connected to the 82188 and the RQ/GT type connected in a daisy-chain to the 8087. The RQ/GT signals form the pattern of three pulses that has been described earlier.

Timing

The timing of wave-forms in the 8087 follows the timing pattern of the 8088/8086, and the differences are in the times rather than the sequences. The table in Figure 6.17 shows the major timing requirements referred to the clock cycle leading or trailing edges. Some of these timings depend on the use of the 8288 or 82188 bus controller.

NOTE: All times in nanoseconds, ns.

Symbol	Parameter	8087 Min	8087 Max	8087-2 Min	8087-2 Max	8087-1 Min	8087-1 Max	Note
TCLCL	CLK cycle period	200	500	125	500	100	500	
TCLCH	CLK low time	118		68		53		
TCHCL	CLK high time	69		44		39		
TCH1CH2	CLK rise time		10		10		15	1
TCL2CL2	CLK fall time		10		10		15	2
TDVCL	set-up time, data in	30		20		15		
TCLDX	hold time, data in	10		10		10		
TRYHCH	set-up time, READY	118		68		53		
TCHRYX	hold time, READY	30		20		5		
TRYLCL	READY inactive to CLK	-8		-8		-10		3
TGVCH	set-up time, RQ/GT	30		15		15		
TCHGX	Hold time, RQ/GT	40		30		20		
TQVCL	set-up time, QS0/QS1	30		30		30		
TCLQX	hold time, QS0/QS1	10		10		5		
TSACH	set-up time, status active	30		30		30		
TSNCL	set-up time, status inactive	30		30		30		
TILIH	Input rise time (not CLK)		20		20		20	
TIHIL	Input fall time (not CLK)		12		12		15	

NOTES: 1. From 1.0 V to 3.5 V levels
2. From 3.5 V to 1.0 V levels
3. Applies only to T2 state (98 ns into T3)

Figure 6.17 *Summary of timing requirements for the 8087.*

Interrupts and exceptions

In the course of carrying out floating-point arithmetic, impossible requirements may arise such as division by zero, square root of a negative number, number overflow or underflow and so on. These conditions are called **exceptions** – not all of them are errors – and each must be catered for. The normal method of coping with exceptions is to return an error or exception value in an interrupt. Such exception values can be distinguished from ordinary floating-point numbers because the floating-point unit excludes numbers with two exponent values, zero and the maximum possible value, from the normal range of floating-point numbers.

Each time an exception occurs, the programmer can make the decision to mask the exception or to trap it and deal with it. The 80387 will carry out default actions on masked exceptions as follows:

1. For an inexact value exception, the result will be rounded to the nearest possible floating-point number. Four methods of rounding can be selected.
2. For masked number underflow (number too small), the 80387 will attempt to produce an approximation by reducing the number of significant figures. If this is not successful, the number is taken to be zero.

3 When a division by zero is masked, the result will be infinity – but with a positive or negative sign depending on the sign bit.
4 For a masked number overflow, the result is again infinity, with the same sign as the overflowing result.
5 When an invalid operation is masked, a 'non-number' is produced – this is referred to as NaN (not a number), and uses a combination of bits impossible to form a genuine floating-point number.

Software

The 8087 keeps an exact copy of the 8088/8086 instruction queue, which is why it is so important to recognize the difference between 8088 and 8086 systems. Each instruction that is intended for the 8087 starts with the binary sequence 11011, which in hex is 1B, denary 27, the ESCAPE code. This is detected by the CPU as the start of a code for the 8087, and will start the sequence of requesting action by the 8087. On 8088/8086 and 80188/80186 systems, software that uses 8087 codes **must** precede each by the byte 9BH (binary 10011011) which is the FWAIT instruction. This ensures that the BUSY line from the 8087 will be tested by the TEST input of the CPU, so avoiding activating another instruction when the 8087 is already working. Software that has been written for the 80287 or 80387 co-processors does not require these wait commands, and they will slow down execution. This is one aspect of code in which there is not perfect compatibility because most software will be written for the most basic processor, retaining the waits which are unnecessary for the later processors.

The ESCAPE code will be decoded and executed by both the 8087 and its host CPU together. What follows is determined by the coding that follows the ESCAPE code. If the instruction makes no reference to memory, no bus activity is needed and the 8087 simply executes the instruction. If the instruction contains a reference to memory (a load or store), then the CPU will calculate the address and perform a read cycle. The 8087 will then capture and save the address in the register. For a read, the 8087 will accept the data word when it becomes available on the data bus. This does **not** require the 8087 to take command of the buses, but if the data consists of more than one word, the 8087 will take bus control, using RQ/GT, and read the rest of the data. On a store from the 8087 to memory, the 8087 ignores the word that is read by the CPU, and takes control of the buses, it then uses the stored address number to perform a write operation on as many bytes of data as it needs to store.

Operations inside the 8087 make use of the data registers as a stack, using Reverse Polish procedures to speed up execution, so that the instructions of the 8087 and the other co-processors in the family are all stack-oriented.

The 80287

The 80287 can be obtained in a variety of speed versions ranging from the 5 MHz chip through 6 MHz, 8 MHz and 10 MHz. The 80287 has internal divider circuits that can divide a clock frequency by 3, so that if the system clock (from the clock chip) is applied to the 80287, a 12 MHz system clock will suit a 5 MHz 80287, a 16 MHz clock will suit the 6 MHz 80287, a 20 MHz clock will suit the 8 MHz 80287 and a 25 MHz system clock will be suitable for the 10 MHz 80287 chip. If the chip is driven from an independent clock generator, such as a separate 8284A, then the rated 80287 frequencies may be used directly.

The pin-out is shown in Figure 6.18, using a

Pin configuration - the # sign indicates active LOW.

1. NC	40. NC
2. NC	39. CKM
3. NC	38. NC
4. NC	37. NC
5. D15	36. PEAK#
6. D14	35. RESET
7. D13	34. NPS1#
8. D12	33. NPS2#
9. V_{CC}	32. CLK
10. V_{SS}	31. CMD1
11. D11	30. V_{SS}
12. D10	29. CMD0
13. NC	28. NPWR#
14. D9	27. NPRD#
15. D8	26. ERROR#
16. D7	25. BUSY#
17. D6	24. PEREQ
18. D5	23. D0
19. D4	22. D1
20. D3	21. D2

Figure 6.18 *Pin allocation for the 80287 which, like the 8087, uses a conventional DIL 40-pin package.*

conventional 40-pin DIL package unlike the 80286. The integration of the 80287 with the 80286 is closer than that of 8087/8088, and the 80287 uses **no address lines** in its connections with the 80286 or with the 80386 with which it is also compatible. An important point about the pin-out is that the pins marked N/C (no connection) must not be connected externally to anything (particularly to earth) since they are likely to be used for internal connections.

The block diagram for the 80287 is identical to that of the 8087, with the division into bus interface unit and numeric execution unit. Like the 8087, the use of the 80287 effectively provides the 80286 with a set of new instructions (just over 50) which deal with floating point arithmetic, trigonometry and various functions. Like the 80286, the 80287 can be operated in real mode or in protected mode, and it will follow the 80286 mode. In real mode, the usual operating mode, the 80287 and 80286 are completely software compatible with the 8087/8086 or 8087/8088 (including the 80186 and 80188). This compatibility also extends to the protected mode. Software written for the 8086/8087 to 8088/8087 will, however, contain FWAIT instructions which are not needed for the 80286/80287, and software will run significantly faster if these instructions are omitted.

Connections

Figure 6.19 shows how the 80287 can be connected in a typical 80286 circuit using the 82C288 bus controller and the 82C284 clock chip for both processors. The bus system is greatly simplified by the lack of address bus for the co-processor, though address decoding is required to provide the NPS1, CMD0 and CMD1 inputs. Figure 6.20 shows the connections to the 80386, for which the address decoding that is required is considerably simpler. In each case the main processor carries out data transfers with memory and generates all the 80287 control signals.

Registers

The registers of the 80287 follow the same pattern as for the 8087 with minor differences. The status word for the 80287 has as its bit 7 the ES (error summary) bit which is set if any unmasked exception bit is set. The control word register has bits 6 and 7 reserved,

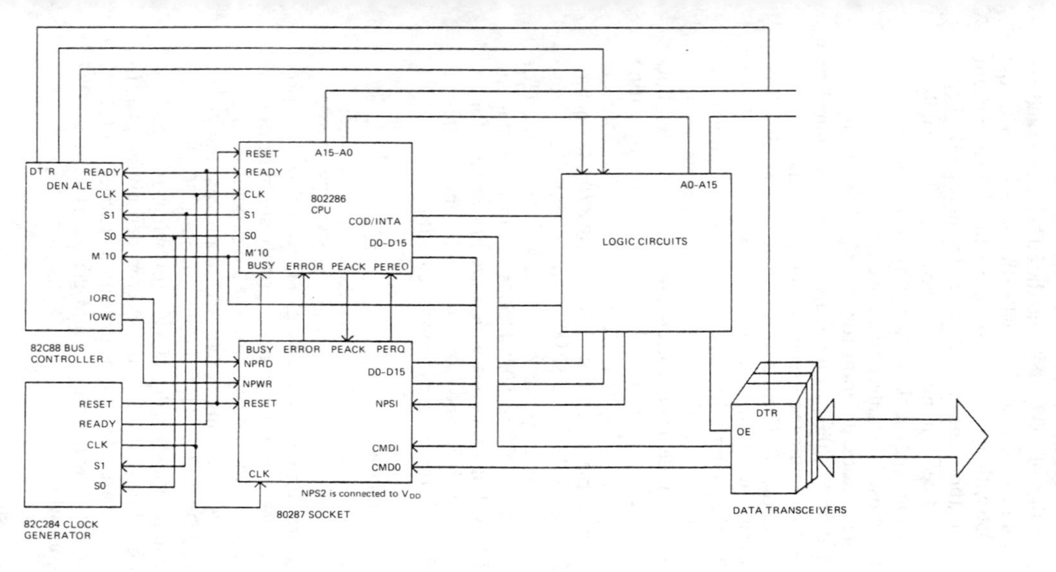

Figure 6.19 *Using the 80287 in a typical 80286 circuit.*

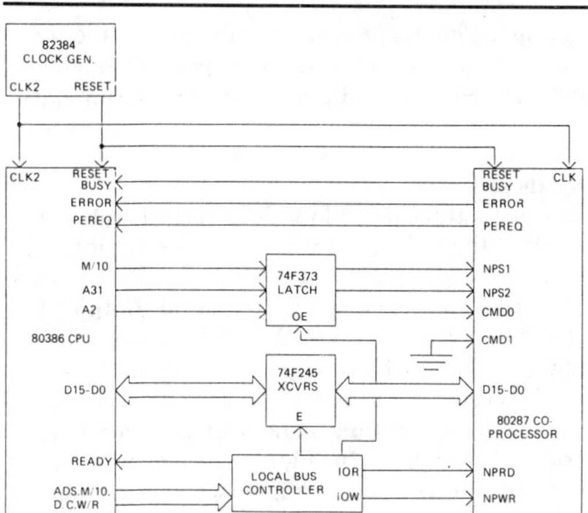

Figure 6.20 *Using the 80287 in a typical 80386DX circuit.*

Figure 6.21 *The control word and status word registers of the 8087.*

omiting the interrupt mask bit of the 8087 (the error signal of the 80287 does not pass through an interrupt controller) – Figure 6.21 shows a summary.

Signals and bus

The signals at the pins of the 80287 closely follow the signals of the 80286, so that shorter descriptions will be used here. DC supplies are V_{cc} of +5 V on pin 9 and earth (ground) on pins 10 and 30. Both of the ground pins **must** be earthed. The V_{cc} pin should be decoupled with a 10 nF capacitor. The CLK input, pin 32, accepts the 33 per cent duty cycle clock from the usual 8284A clock generator chip to provide the basic 80287 timing. In addition, there is a CKM input, pin 39, which determines whether or not the CLK input is divided. With CKM input HIGH, the clock frequency is used directly; with CKM LOW the clock input is divided by three as described earlier. This pin should be tied to V_{cc} or V_{ss} and never left floating. The voltage must be applied to CKM at least 20 CLK cycles before RESET goes LOW.

The RESET input on pin 35, active HIGH, causes the co-processor to terminate any current instruction. The signal must be active for at least four cycles of the system clock, and RESET must not be released (HIGH to LOW change) less than 50 μs after V_{cc} and the CLK signal have settled to their specified values.

The other inputs are for synchronization of the data transfers, and correspond to 80286 outputs. The PEACK input, pin 36, is active LOW and is an acknowledge for the PEREQ (processor extension data channel operand transfer request). If no transfers are pending and the PEREQ signal is given out, the PEACK will cause the PEREQ to be reset. The PEACK signal can be asynchronous to the CLK signals. The PEACK pin on the 80287 is connected to the corresponding pin of the 80286, but for use with the 80386, the PEACK input is connected to V_{cc} by way of a 20 Ω pull-up resistor.

The NPRD input on pin 27 is the numeric processor read enable, which permits data to be read **from** the 80287. This input is active LOW, and would normally be obtained from the IORC output of the 82288 bus controller, or whatever other bus controller is used. The same comments apply to the NPWR input on pin 28, also active LOW. This latter pin would be connected to the IOWC output of the bus controller, and enables writes to the co-processor.

The numeric processor select pins are labelled NPS1 and NPS2 on pins 34 and 33 respectively. NPS1 is active LOW and NPS2 is active HIGH, and both inputs must be TRUE together to select the co-processor chip. The pins can be driven by asynchronous signals, and no transfer of data to or from the co-processor will be possible if both selects are not at their ON levels of LOW for NPS1 and HIGH for NPS2. These pins will be driven by the address selection logic of the main processor, and the circuit examples show this done for the 80286 (with NPS2 held HIGH) and for the 80386 (with both driven).

The other inputs are the CMD0 and CMD1 command lines on pins 29 and 31 respectively. These are connected, by way of latching, to the A1 and A2 address lines of the main processor in order to establish the correct hardware address.

Of the outputs, BUSY on pin 25 is active LOW and will be connected to the BUSY pin of the main processor. BUSY remains LOW while the 80287 is executing a command, and its connection to the BUSY input of the 80286 or 80386 will ensure that no further instruction can be passed to the 80287 until this signal level reverts to HIGH. This provides an automatic way in which the CPU will check whether the 80287 is ready for input, avoiding the need for software to contain the FWAIT instruction for this purpose, as is necessary for the 8087.

The ERROR output, active LOW, is on pin 26 and connects to the ERROR pin of the main processor. This signal goes LOW to signal that the error status bit in the status word has been set, and that an unmasked error exists. The other output is PEREQ, active HIGH, on pin 24, which connects to the pin of the same title on the main processor. This signal goes HIGH to signal that the 80287 is ready to transmit data, having completed its numerical task.

The remaining pins are data lines used for both input and output. These form a 16-bit bidirectional bus whose DT inputs and output can be asynchronous. The connections to the main processor can be direct (for the 80286) or by way of a transceiver (80386) – note that when the 80287 is used with the 80386, only the lower 16 data lines are used.

Interfacing

The interfacing of the 80287 is very different from that of the 8087, since with no address pins being used on the 80287, the addressing is as a port. This uses the reserved port addresses 00F8H, 00FAH and 00FCH (port addresses from 00F8H to 00FFH are reserved). If the 82288 bus control unit is present, it can identify port transfers and provide IORC and IOWC signals which connect to the NPRD and NPWR pins of the 80287. Any bus drivers on the local 80286 bus must be disabled during a read of the 80287.

The differences in the interfacing of 80286 and 80386 are fairly minor. For the 80386 interfacing, the PEACK pin is maintained HIGH because its use is not required. This pin is used to keep track of the number of words of operand, a task that is unnecessary when the 80386 is used. In addition, because the 80386 local bus is clocked at a very high speed, a local bus controller is needed to generate the correct read and write timings for the 80287, and the chip-select signals have to be generated as well. Since the 80386 uses port addresses 800000F8H to 800000FFH to work with the 80287, the chip address signals are easy to decode using A31 and M/IO. When the 80287 is connected to the 80386, the CPU will **automatically** make use of 16-bit transfer cycles.

Interrupts

The 80286/80287 connection does not involve the use of hardware interrupts, and any error (exception) is notified by way of the ERROR signal line. Two interrupt numbers are reserved in the 80286 for use with 80287 exceptions, and these were listed in Figure 4.25.

Error handling is the main source of incompatibilities between 8087 and 80287 use, since the error signal from the 80287 does not pass through an interrupt controller, unlike the INT signal of the 8087. Any software that contains instructions aimed at an interrupt controller will cause problems when run on the 80286/80287 combination. The vector for interrupt 16 must point to the handler routine for co-processor errors.

Among more minor points:

- The floating-point instruction address that is saved in the register of the 80287 includes any leading prefixes **before** the ESCAPE code – these prefixes are not present in the 8087 register code.

- When run in protected mode, the 80286/80287 combination uses a different format for instruction and operand pointers – nothing in the 8087 corresponds to this, but it is hardly likely to be a worry as far as compatibility is concerned.
- Interrupt 7 occurs when a co-processor instruction is executed with TS, EM or MSW=1. If TS or MSW=1 and WAIT is used, the exception 7 will also be generated. An interrupt handler will be needed to deal with this.
- There will be an interrupt 9 if the second (or later) words of a floating-point operand occur outside the size limits of a segment. If the starting address is outside the segment, interrupt 13 is generated. An interrupt controller is needed to deal with these problems.

Timing

The timing of wave-forms in the 80287 follows the timing pattern of the 80286. Figure 6.22 shows the timing of a data transfer which has been initiated by the 80286, starting with changes in the select lines

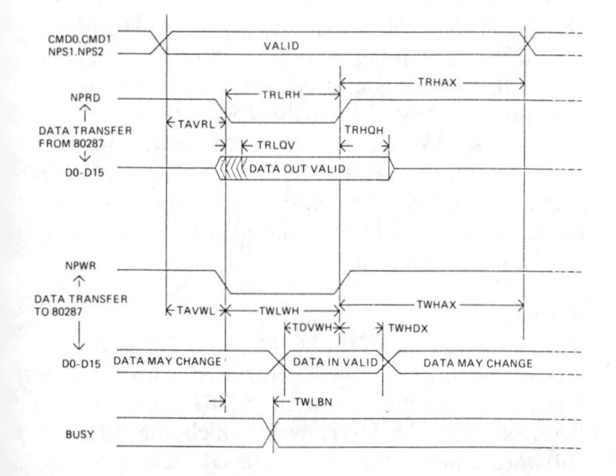

Figure 6.22 *Timing of a data transfer, which is initiated by the 80286. Both directors of data transfer are illustrated.*

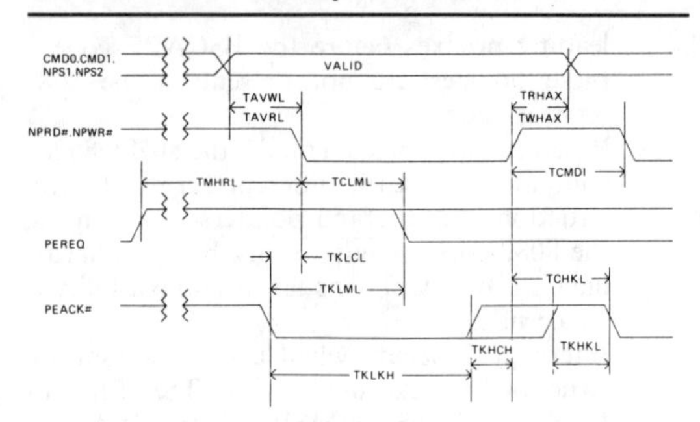

Figure 6.23 *Data channel timing, which is initiated by the 80287.*

CMD and NPS. The changes in the select voltages allow the NPRD and NPWR inputs to follow, and the drawings show the timing of either a read or a write produced in this way. The data channel timing is initiated by the 80287, and Figure 6.23 shows this timing pattern, which makes use of the PEREQ, PEACK form of handshaking.

The 80286 and 80287 work together with numerical instructions, with the 80287 controlling the data bus only when a transfer is made to the 80286. Like the 8087, the 80287 contains the same instruction queue as the 80286. When a command that starts with the ESCAPE code is encountered by the 80286, the BUSY pin voltage is tested, and the 80286 will wait if this line indicates that the 80287 is working. When this signal is cleared, the data is transferred and the 80287 works on the ESCAPE instruction while the 80286 continues with the following instruction (assuming that this is not also an ESCAPE sequence). If 8087 software is being used, the presence of the FWAIT instruction will cause a delay but has no other effect.

Software

The programming of the 80286 and 80287 is done as if only one processor were being used, with the same ESCAPE sequence in each 80287 code as was described for the 8087, with which the 80287 is software compatible. As has been mentioned, no FWAIT needs to be inserted **before** an 80287 instruction, but **must** be used **after** any 80287 store to memory or load from memory (exceptions are

FSTSW, FSTCW, FLDENV and FRSTOR). This gives the 80287 time to alter memory before the same memory can be used by the 80286.

The 80387

The 80387 can be obtained in a variety of speed versions ranging from the 16 MHz chip through 20 MHz to 25 MHz. The 80387 has internal divider circuits that can divide the clock frequency by 2, so that if the system clock (from the clock chip) is applied to the 80387, the clock will be synchronous to that of the 80386 with which the 80387 is being used. The chip can be used in asynchronous mode, but this refers to data transfers, control logic and floating-point operations only; the bus control logic is always synchronous with the 80386DX CPU.

The pin-out is shown in Figure 6.24, using a dual-row pin–grid system. No address lines are used, because like the 80287, the 80387 uses port (I/O) addresses of the 80386. Like the 80386, the 80387 uses a large number of V_{cc} and V_{ss} connections, all of which must be correctly connected. Pin K9 **must** not be connected, and pins K3 and L9 must be connected to V_{cc}.

The block diagram for the 80387 is illustrated in Figure 6.25, with the division in this case into bus control logic, data interface and control unit and floating point unit. Like the 80287, the use of the 80387 effectively provides the 80386 with a set of new instructions (just over 70) which deal with floating-point arithmetic, trigonometry and various functions. The 80387 can be operated independently of the mode of the 80386, whether this is real mode, protected mode, or virtual 8086 mode. Access to memory is handled by the 80386, and the 80387 operates on values and instructions that have been passed to it along the 32 data lines. When real or virtual-8086 mode is used by the 80386, the combination of 80386 and 80387 will run software written for 8088/8087, 8086/8087, or 80286/80287 (real mode). A form of pipelining is used within the 80387, in that a numeric instruction can be processed while the commands and data for the next instruction are being transferred.

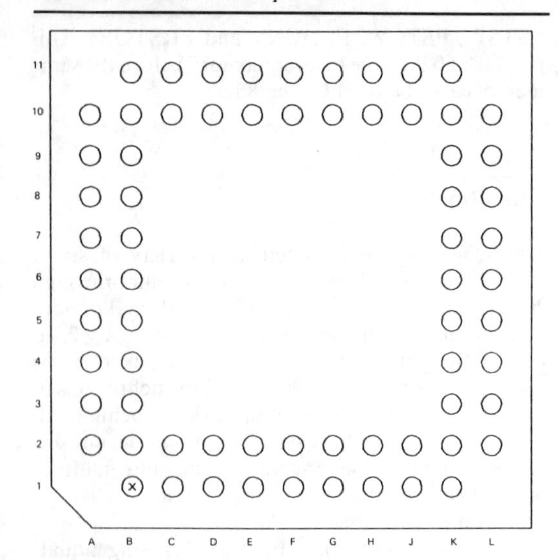

The hashmark (#) is used to identify pins whose action is active LOW.

Pin	Signal	Pin	Signal	Pin	Signal
A2	D9	C11	V_{SS}	J10	V_{SS}
A3	D11	D1	D5	J11	CKM
A4	D12	D2	D4	K1	PEREQ
A5	D14	D10	D24	K2	BUSY#
A6	V_{CC}	D11	D25	K3	Tie to 1
A7	D16	E1	V_{CC}	K4	W/R#
A8	D18	E2	V_{SS}	K5	V_{CC}
A9	V_{CC}	E10	D26	K6	NPS2
A10	D21	E11	D27	K7	ADS#
B1	D8	F1	V_{CC}	K8	READY#
B2	V_{SS}	F2	V_{SS}	K9	NC
B3	D10	F10	V_{CC}	K10	CPUCLK2
B4	V_{CC}	F11	V_{SS}	K11	NUMCLK2
B5	D13	G1	D3	L2	ERROR#
B6	D15	G2	D2	L3	READYO#
B7	V_{SS}	G10	D28	L4	STEN
B8	D17	G11	D29	L5	V_{SS}
B9	D19	H1	D1	L6	NPS1#
B10	D20	H2	D0	L7	V_{CC}
B11	D22	H10	D30	L8	CMD0#
C1	D7	H11	D31	L9	Tie to 1
C2	D6	J1	V_{SS}	L10	RESETIN
C10	D23	J2	V_{CC}		

Figure 6.24 *The pin diagram (a) and pin listing (b) for the 80387.*

Connections

Figure 6.26 shows how the 80387 can be connected in a typical 80386 circuit using a wait-state generator. The bus system is as simple as that of the 80286/80287 system, with virtually all of the pins of the 80387 connecting directly with pins of the 80386 or associated clock chips.

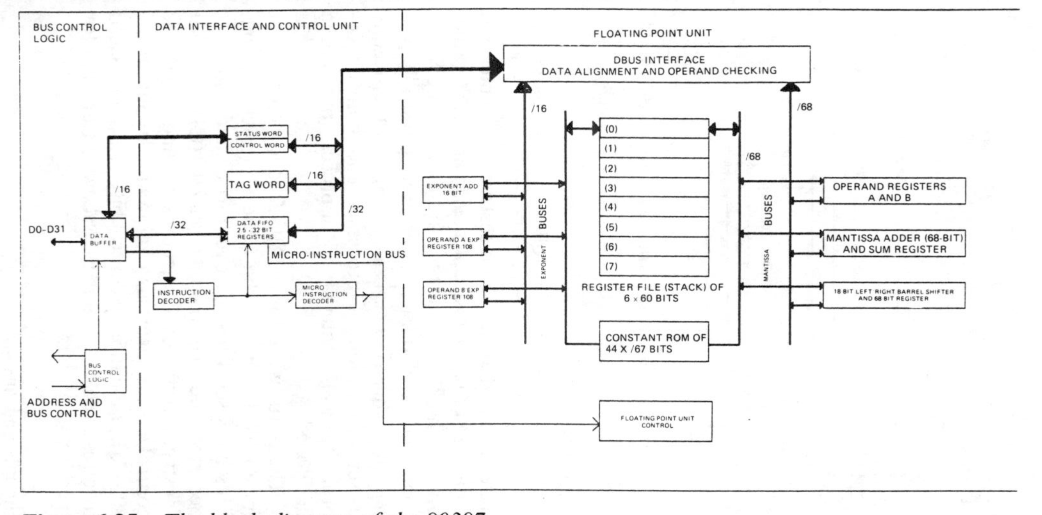

Figure 6.25 *The block diagram of the 80387.*

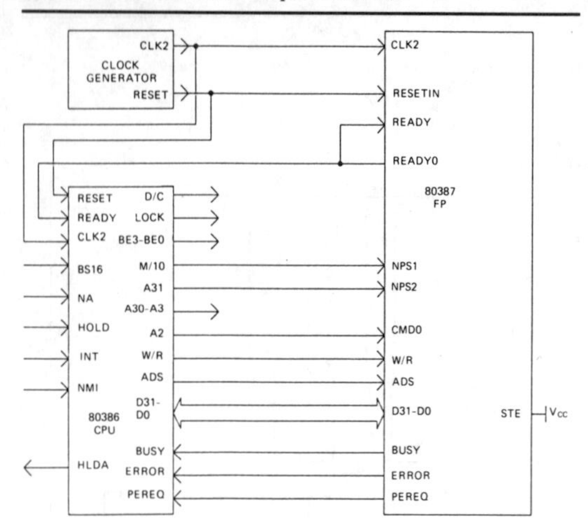

Figure 6.26 *Connecting the 80387 to the 80386DX.*

Registers

The registers of the 80387 follow the same pattern as for the 80287 with minor differences. The main difference is that the 48-bit instruction pointer and data pointer registers are **located in the 80386** rather than in the 80387, but they are used by the 80387 commands as if these registers were physically present in the 80387. This is an example of the steadily closer integration of processor and numeric co-processor as the Intel family of processors has been developed.

The status word for the 80387 follows the layout for 80287, but has for bit 6 the SF (stack flag) bit. This bit is set if a stack overflow or underflow has occurred, and when this bit is set, bit 9 (C1) will be set for stack overflow and reset for stack underflow. The control word register also follows the layout for the 80287, but the infinity control bit has been replaced by a reserved bit. Figure 6.27 shows a summary for these two registers.

Signals and bus

The signals at the pins of the 80387 closely follow the signals of the 80386. DC supplies are V_{cc} of +5 V on pins A6, A9, B4, E1, F1, F10, J2, K5 and L7, and earth/ground on pins B2, B7, C11, E2, F2, F11, J1,

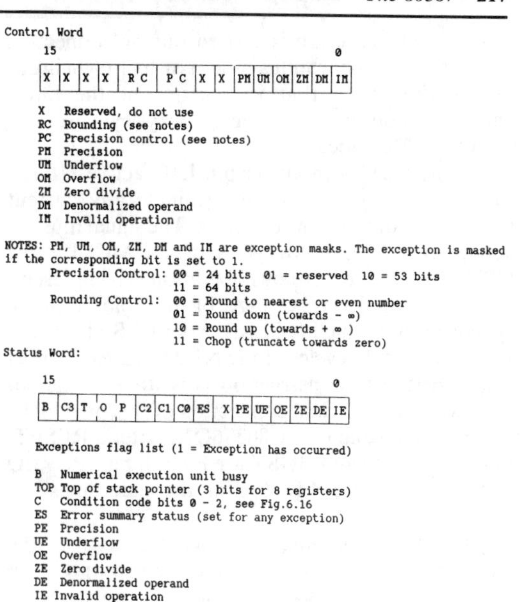

Figure 6.27 *The control word and status word registers of the 80287.*

J10 and L5. All of these pins **must** be appropriately connected.

There are three pins connected with clocking, CPUCLK2 (pin K10), NUMCLK2 (pin K11) and CKM (pin J11). The CPUCLK2 is an input for clock pulses from the chip that also provides the CLK2 signal to the 80386. This input will time the bus control logic and when synchronous mode is selected with CKM, this input times also the data interface, control unit and the floating-point unit of the 80387. Signal levels are MOS, and the input is internally divided by 2 as for the 80386 CLK2 input.

The NUMCLK2 input is intended for use with asynchronous mode, when this input will be the clock for the data interface, control unit and floating-point unit. The frequency of this clock can be more or less than that of the CPU clock, but only within a limited range, down to $\frac{5}{8}$ of the CPUCLK2 frequency, or up to 7.5 times of the CLK2 frequency. The pin, whose inputs are TTL signals, is ignored when synchronous mode is selected.

The mode selection is carried out by connecting CKM to 0 (asynchronous mode) or to 1 (synchronous mode). Note that irrespective of the mode chosen, the bus control logic is always synchronous with the CPU clock.

The RESETIN input on pin L10, active HIGH, causes the co-processor to terminate any current instruction and become dormant. The signal must be active for at least 40 cycles of NUMCLK2 (or CPUCLK2 if synchronous operation is being used). The High to LOW change of this pin **must** be synchronous with CPUCLK2. After RESETIN goes LOW at least 50 cycles of the NUMCL2 clock must elapse before any instruction is written to the co-processor. The normal connection of this pin will be to the RESET pin of the 80386DX. After a RESET, pins READYO and BUSY are HIGH, pins PEREQ and ERROR are LOW, and the data lines are floating.

The other inputs are for synchronization of the data transfers, and correspond to 80386 outputs. No PEACK input is used (see the comments on the 80287 used with the 80396).

The chip and port select pins are STEN, NPS1, NPS2 and CMDO. Of these, STEN (status enable) operates as a chip select. When STEN is LOW the BUSY, PEREQ, ERROR and READYO output are floated, and other chip select signals are inactivated. The STEN input is intended primarily for testing and for many applications of the 80387 this pin can be wired to the V_{cc} voltage through a 20 kΩ pull-up resistor.

The main select inputs are NPS1 and NPS2. Both are active LOW and are on pins L6 and K6 respectively. The NPS1 pin will usually be connected to the M/IO pin of the 80386, so that this chip-select signal is active when the 80386 is making access to a port. The NPS2 pin will then be connected to the A31 pin of the 80386, so that only a port address in the range 800000F8H to 800000FFH can be used. The other select is CMDO, active low, and on pin L8. This input should be LOW for an opcode, HIGH for data on a write cycle. During a read cycle it is LOW for control or status register and HIGH for a data register being read. The normal connection of CMDO is to the A2 pin of the 80386DX.

The other inputs are W/R, ADS and READY.

The W/R pin is K4 and is connected to the 80386 pin of the same name. This signal is HIGH for a write cycle and LOW for a read cycle, so that the 80387 is informed of the bus cycle that the 80386 is carrying out on its behalf. Voltages on this pin will be ignored when the 80387 is not selected by STEN, NPS1 or NPS2.

The address strobe ADS is on pin K7, active LOW. This pin is connected to the ADS pin of the 80386 and its voltage goes LOW to indicate that the bus-control logic of the 80387 may sample the signals on W/R, NPS1 and NPS2. The READY input, active LOW, is on pin K8 and is used to signal to the 80387 when a bus cycle is to be ended. The READY voltage can be kept HIGH to add wait states to a bus cycle, as on the 80386.

Of the outputs, BUSY on pin K2 is active LOW and will be connected to the BUSY pin of the main processor. BUSY remains LOW while the 80387 is executing a command, and its connection to the BUSY input of the 80386 will ensure that no further instruction can be passed to the 80387 until this signals level reverts to HIGH. This provides the automatic way in which the CPU will check whether the 80387 is ready for input, avoiding the need for software to contain the FWAIT instruction for this purpose, as is necessary for the 8087.

The ERROR output, active LOW, is on pin L2 and connects to the ERROR pin of the main processor. This signal goes LOW to signal that the error status bit in the status word has been set, and that an unmasked error exists. The signal can be deactivated only by the instructions FNINIT, FNCLEX, FNSTENV or FNSAVE. PEREQ, active HIGH, on pin K1, connects to the pin of the same title on the main processor. This signal goes HIGH to signal that the 80387 is ready to transmit or receive data.

The other output is READYO which is used mainly when no wait states are needed. The pin voltage goes high after two clocks on a write cycle and three clocks on a read cycle and when no wait states are used, connecting this voltage to the ARDY input of the CPU will terminate the cycles. When wait states are used, as is often the case with the faster processor/co-processor combinations, the READYO signals can be combined with wait signals from memory by using a wait state generator. The READYO **must** be connected in one of two possible ways:

1 connected directly or through an OR-gate to the ARDY inputs of the 80386DX and the 80387;
2 used as one input to a wait-state generator.

The remaining pins are the 32 data lines used for both input and output. These form a bidirectional bus which is normally connected directly to that of the 80386DX.

Interfacing

The interfacing of the 80387 is very similar to that of the 80287, addressing the co-processor as a port. The important difference is that the port address for the 80387 is not in the normal range of I/O addresses for the 80386 chip, using the A31 line HIGH in each access. This makes it unnecessary to use any reserved port addresses, since no normal port would be mapped in this range. The PEACK input of the 80287 is not used on the 80387.

Interrupts

The 80386/80387 connection does not involve the use of hardware interrupts, and any error (exception) is notified by way of the ERROR signal line. Four interrupt numbers are reserved in the 80286 for use with 80387 exceptions, and these are as listed previously for the 80287.

Error handling is the main source of incompatibilities between 8087, 80287 and 80387 use. The 80387 was designed in the light of the final IEEE standards, so that there have been considerable changes in the architecture of the chip as compared to the 80287, many of which increase the performance of the chip and eliminate the need for software to be used to implement the IEEE standard. The differences are specialized, and mathematicians who are programming the 80387 should consult the Intel data for details. The most noticeable change is that no infinity control is needed, because the 80387 supports only the affine closure for infinity (see the description earlier under the 8087 heading). The main differences in operation affect exception handling, and this is also of a rather specialized nature.

Timing

The timing of wave-forms in the 80387 follows the timing pattern of the 80386. A new bus cycle is started by the 80386 when the ADS voltage is taken LOW. If the chip select pins of the 80387 are all active at that time, this is recognized as the start of an 80387 bus cycle. The BS16 input **must not be used** during cycles when A31 is active, signalling an access to the 80387. While ADS is active, the WR input is examined by the 80387 to determine when this will be a read or write cycle, and CMDO will be used to distinguish an opcode from operand or control register from status register transfer.

The 80387 allows either pipelined or non-pipelined operation. When ADS is put LOW with no other 80387 bus cycle in progress, the cycle will be non-pipelined. When an 80387 action is under way and ADS is put low again, this allows for pipelining so that valid signals for next address and next control signals can be read. This matches the pipelining of the 80386 itself.

Like the 80386, the timing of the 80387 is referred to bus states, and the bus state diagram is illustrated in Figure 6.28. The Ti (Idle) and Trs (READY-sensitive) states are used when non-pipelined action is required, and the Tp state when pipelining is in use. After a RESET, the bus logic will be in the idle state, and this is also the state used following a non-pipelined cycle and also following the end of pipelining of a number of pipelined cycles. The READY-

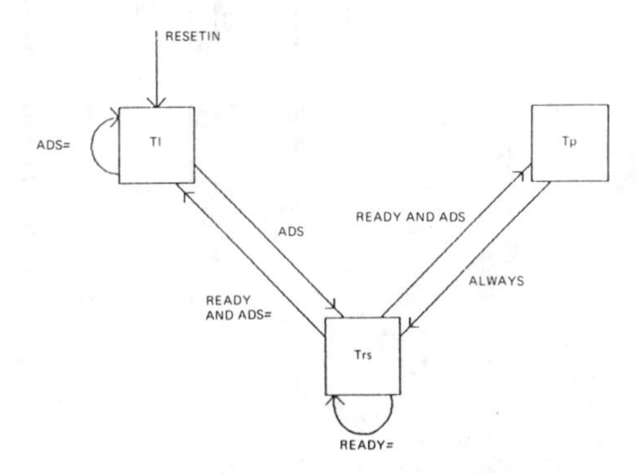

Figure 6.28 *A bus state diagram for the 80387. Ti=Idle, Trs= Ready, Tp=pipeline.*

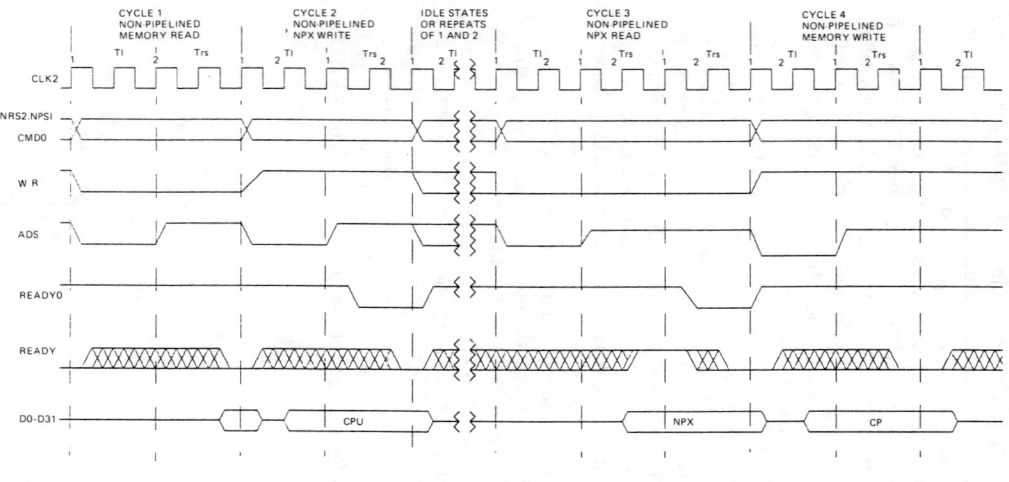

Figure 6.29 *Timing of non-pipelined read and write cycles for the 80387.*

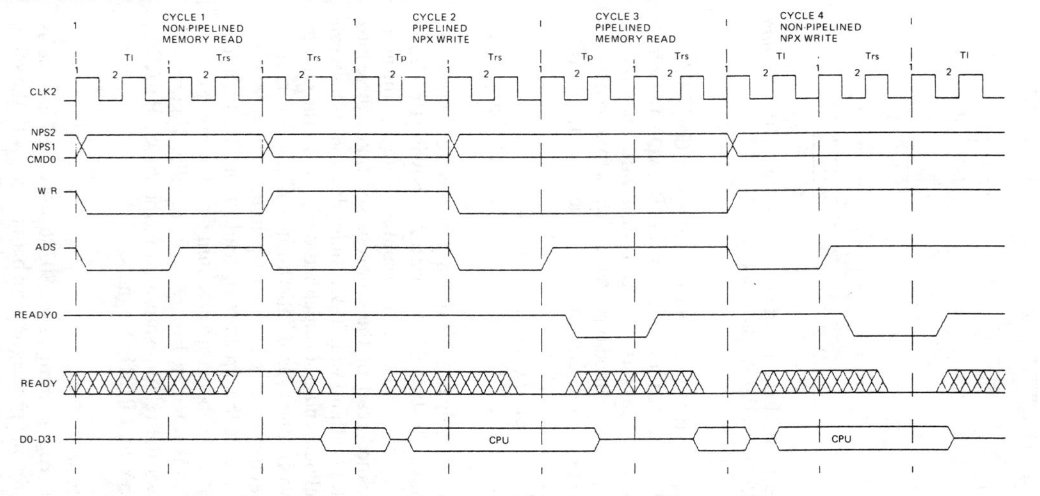

Figure 6.30 *Timing of the fastest possible transitions between pipelined and non-pipelined cycles.*

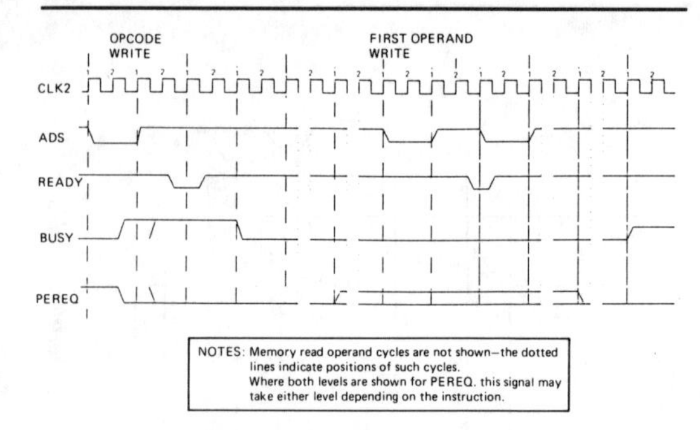

Figure 6.31 *The ADS, READY, BUSY and PEREQ signal relationships.*

sensitive stage is the one in which asserting READY will terminate the cycle, allowing READY to be held so as to repeat this state, causing a wait. The Tp state is used in pipelined bus cycles, always followed by the Trs state (note this does not have to wait for ADS, since the next input has already occurred). All of the activities of the 80387 bus interface occur either during the Trs state or in the course of changing to that state.

The timing of typical non-pipelined read and write cycles is illustrated in Figure 6.29, showing the CLK2 signal as the master timing source. Note that the non-pipelined read of the co-processor takes **three** bus cycles (with no waits). The main differences between pipelined and non-pipelined cycles lie in the change between one type of operation and the other, since the active state is Trs, and pipelining simply involves the use of the Tp state instead of the Ti state. The transitions between pipelining and non-pipelining are illustrated in Figure 6.30. Finally, Figure 6.31 shows the relationship among the ADS, READY, BUSY and PEREQ signals.

Software

The programming of the 80386 and 80387 is done as if only one processor were being used, with the same ESCAPE sequence in each 80387 code as was described for the 8087, with which the 80387 is software compatible. As has been mentioned, the redesign of the 80387 to incorporate the final IEEE standard means that older mathematical software

The hashmark (#) is used to identify pins whose action is active LOW.

Pin	Signal	Pin	Signal	Pin	Signal	Pin	Signal
1	NC	18	NC	35	ERROR#	52	NC
2	D7	19	D0	36	BUSY#	53	NUMCLK2
3	D6	20	D1	37	V_{CC}	54	CPUCLK2
4	V_{CC}	21	V_{SS}	38	V_{SS}	55	V_{SS}
5	V_{SS}	22	V_{CC}	39	V_{CC}	56	PEREQ
6	D5	23	D2	40	STEN	57	READYO#
7	D4	24	D8	41	W/R#	58	V_{CC}
8	D3	25	V_{SS}	42	V_{SS}	59	CKM
9	V_{CC}	26	V_{CC}	43	V_{CC}	60	V_{SS}
10	NC	27	V_{SS}	44	NPS1#	61	V_{SS}
11	D15	28	D9	45	NPS2	62	V_{CC}
12	D14	29	D10	46	V_{CC}	63	V_{SS}
13	V_{CC}	30	D11	47	ADS#	64	V_{CC}
14	V_{SS}	31	V_{CC}	48	CMD0#	65	NC
15	D13	32	V_{SS}	49	READY#	66	V_{SS}
16	D12	33	V_{CC}	50	V_{CC}	67	NC
17	NC	34	V_{SS}	51	RESETIN	68	NC

Figure 6.32 *The top-view pin diagram (a) and pin assignment (b) for the 80387SX co-processor.*

designed for the 8087 or 80287 should be carefully examined to check that the changed exception handling of the 80387 will be correctly used.

The 80387SX

The 80387SX was originally specified in a 16 MHz version only, and at the time of writing, some information was still on a tentative basis. The pin-out is shown in Figure 6.32, using a PLCC format with 68 pins. As for the 80387, there are many V_{cc} and V_{ss} pins, all of which **must** be connected, and eight NC pins, **none** of which must be connected to anything.

The remainder of the 80386SX operation is so similar to that of the 80387 chip (which may be renamed as 80387DX in line with the CPU chips) that only the essential differences need be covered here. Only 16 data pins are needed for the SX version of the chip, and the decoding for NPS2 is provided by the A23 address pin to ensure that the port addresses that are used are 8000F8H to 8000FEH. The address 8000FEH is treated as identical to 8000FCH by the SX chip. The address 8000F8H is used when the 80386SX is writing a command, and 8000FCH or 8000FEH is used when the 80386SX is writing or reading data.

7 Support devices

The support devices that are the subject of this section are the essential chips that allow the CPUs to form a rudimentary system. The numeric co-processors have been treated as being part of the CPU bus in the previous section. Because so many of the support chips have been mentioned in the course of describing CPU operations, only the essential points have been summarized in this section, keeping as far as possible to the scheme of describing one family of devices at a time. Where a support chip is programmable, details of its software aspects have been omitted or summarized, since the main thrust of this book concerns the CPU chips. Details of timing wave-forms are also omitted, and the designer of a system using these chips should refer to the appropriate Intel manuals for such information. Some of these support chips are comparable in complexity with the CPUs that they serve, and a full treatment would be impossible in the space of this book. The simpler devices are described in more detail since they are more likely to be used in controller applications; whereas the devices that are intended for mainly computing applications are used in circuits which are stereotyped and well known.

Since the main task of this section is to note the essential bus support chips, no descriptions of other support devices such as I/O, disk controllers, graphics co-processors, etc. are included.

Clock chips

The original clock generator for the 8088/8086 CPUs was the bipolar 8284, but this has been superseded for computing purposes by the CMOS version, the 82C84A. The bipolar version is still available for use in controllers. This clock generator and driver chip, which exists in 5 MHz (82C84A-5) and 8 MHz (82C84A) versions. This chip also provides for the

```
The # sign means that the pin is active LOW

Pin  Signal       Pin  Signal
 1   CSYNC        18   Vcc
 2   PCLK         17   X1
 3   AEN1#        16   X2
 4   RDY1         15   ASYNC#
 5   READY        14   EFI
 6   RDY2         13   F/C#
 7   AEN2#        12   OSC
 8   CLK          11   RES#
 9   GND          10   RESET
```

Figure 7.1 *Pin assignments for the 82C84A, using a conventional 18-pin DIL package.*

Control Pin	Action 0	Action 1
F/C#	Crystal drive	External clock
RES#	Reset	Normal
RDY1 RDY2	Bus not ready	Bus ready
AEN1# AEN2#	Address enabled	Address disabled
ASYNCH#	2 stage ready synchronization	1 stage ready synchronization

Figure 7.2 *The control inputs of the 82C84A.*

synchronization of the READY signal and provision for RESET, and its pin-out, using a conventional 18-pin DIL format, is shown in Figure 7.1. The chip contains the oscillator circuit which can be controlled by an external crystal, a divide by three counter, and the logic for READY synchronization and RESET timing.

Seven control inputs are used, and their actions are summarized in the table of Figure 7.2. The X1 and X2 pins are used for crystal connection or external input – if external input is used X2 is unconnected and the X1 pin is connected to V_{cc}. External clock inputs are connected to EFI (with F/C HIGH), but when crystal control is used, the EFI pin should be connected to V_{cc} or earth/ground. The CSYNC input, active HIGH, allows the 82C84 to be synchronized to another clock. This input acts by resetting the internal counters when CSYNC is HIGH, and enabling counting when the pin voltage is LOW. If no other clock exists, this input should be tied to earth/ground.

The oscillating circuit for crystal control is shown in Figure 7.3, using two low-loss capacitors of equal value. The series combination of these values (equivalent to half the value of one capacitor), plus

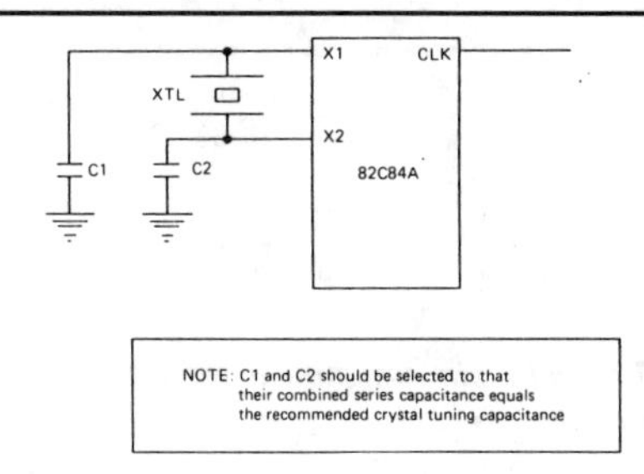

Figure 7.3 *The crystal connections for the 82C84.*

an allowance for stray capacitance, should equal the load capacitance as recommended by the manufacturer of the crystal. The crystal frequency should be three times the final CLK frequency.

The CLK output has a duty cycle of $\frac{1}{3}$, and a frequency which is $\frac{1}{3}$ of the crystal frequency (or of the frequency externally input at EFI). This signal is suitable for driving MOS devices. The PCLK output is intended for clocking slower peripherals, and has a frequency which is half of that of the CLK frequency, along with a 50 per cent duty cycle. Signal levels are TTL. A third output is on the OSC pin, and is a TTL level signal at oscillator frequency.

The READY output can be connected to the READY pin of the CPU. This signal is active HIGH, and is the result of synchronizing the RDY1 or RDY2 inputs. The RESET output is active HIGH, and is established after a delay caused by an RC timing network at the RES input, so providing the correct timing of the RESET voltage when the power is first applied to the chips of a system.

The 82C284

This is the recommended clock generator for the 80286 CPU and its peripherals. This clock generator and driver chip exists in 8 MHz, 10 MHz and 12 MHz versions, which are distinguished by -8, -10 and -12 suffixes respectively. This chip also provides for the synchronization of the READY signal to the 80286, and a RESET output. The chip can use either a conventional 18-pin DIL casing, Figure 7.4, or a 20-pin PLCC, Figure 7.5.

The # sign means that the pin is active LOW

Pin	Signal	Pin	Signal
1	ARDY#	18	V_{CC}
2	SRDY#	17	ARDYEN#
3	SRDYEN#	16	S1#
4	READY#	15	S0#
5	EFI	14	NC
6	F/C#	13	PCLK
7	X1	12	RESET
8	X2	11	RES#
9	GND	10	CLK

Figure 7.4 *The pin assignment for the 20-pin DIL package of the 82C284.*

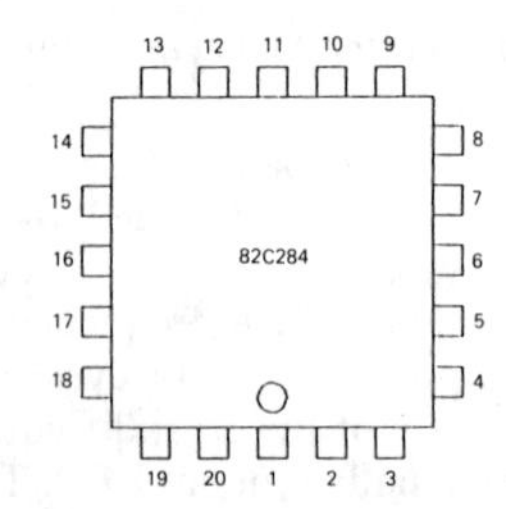

The # sign means that the pin is active LOW

Pin	Signal	Pin	Signal
1	NC	11	NC
2	ARDY#	12	CLK
3	SRDY#	13	RES#
4	SRDYEN#	14	RESET
5	READY#	15	PCLK
6	EFI	16	GND
7	F/C#	17	S0#
8	X1	18	S1#
9	X2	19	ARDYEN#
10	GND	20	V_{CC}

Figure 7.5 *The pin diagram (a) and assignment (b) for the PLCC version of the 82C284.*

Control input	Action
F/C#	0=Use internal oscillator, 1=Use EFI input
RES	Generate system reset, active LOW
ARDY	Terminate bus cycle, asynchronous, active LOW
SRDY	Terminate bus cycle, synchronous, active LOW
ARDYEN	Selects ARDY, active LOW
SRDYEN	Selects SRDY, active LOW

Figure 7.6 *The six control inputs for the 82C284 chip.*

Six control inputs are used, and their actions are summarized in the table of Figure 7.6. The X1 and X2 pins are used for crystal connection – if external input is used X2 is unconnected and the X1 pin is connected to V_{cc}. External clock inputs are connected to EFI (with F/C HIGH), but when crystal control is used, the EFI pin should be connected to V_{cc} or earth/ground.

The oscillating circuit for crystal control uses the circuit previously illustrated in Figure 7.3, using two low-loss capacitors of equal value ranging from 60 pF for crystal frequencies of 1–8 MHz, 25 pF for 8–20 MHz and 15 pF for frequencies higher than 20 MHz. The crystal frequency is the same as the CLK frequency, and should be twice the frequency of the internal processor clock (the 80286 will divide its clock input frequency by two). The CLK output has a duty cycle of 50 per cent. At the higher CLK frequencies the CLK line should be treated as a transmission line and be correctly terminated, usually by adding a series resistor which will bring up the output impedance of the chip to the impedance level of the line. Resistor values in the range 10R to 74R can be used. The PCLK output is intended for clocking slower peripherals, and has a frequency which is half of that of the CLK frequency, also with a 50 per cent duty cycle.

The READY output can be connected to the READY pin of the CPU. This signal is active HIGH, and is the result of the control inputs, all of which are active LOW. Two sources of READY can be used, synchronous at the SRDY input or asynchronous at the ARDY input. Each of these inputs has an enable input (SRDYEN or ARDYEN) so that the source of the READY input can be selected, and one of these inputs would be selected by using an address decoder to act on its enable terminal. The READY output becomes active (LOW) at a sampling time when both SRDY and SRDYEN are both active, or when both ARDY and ARDYEN are active. The READY output will remain active for at least two CLK cycles (one processor cycle). The output is of the open-drain type and should be connected to a pull-up resistor, typically 20 kΩ.

The READY signal must be forced HIGH (inactive) at the start of a bus cycle, and this is done by using the S0 or S1 inputs (both active LOW), which have internal pull-up resistors. Making either of these inputs active will float the READY output. Figure 7.7 illustrates the synchronous READY (SRDY) action with reference to the Ts, Tc and Ti states of the processor (see Section 5). The asynchronous operation is summarized in Figure 7.8.

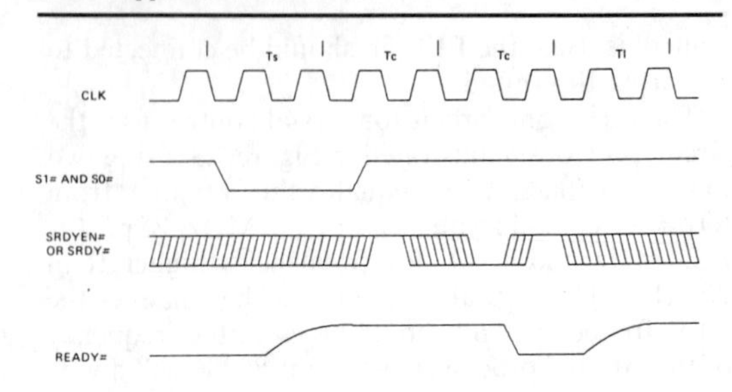

Figure 7.7 *The SRDY output used for synchronous READY timing. ARDYEN must be held HIGH and for another READY, both SRDYEN and SRDY must be taken LOW.*

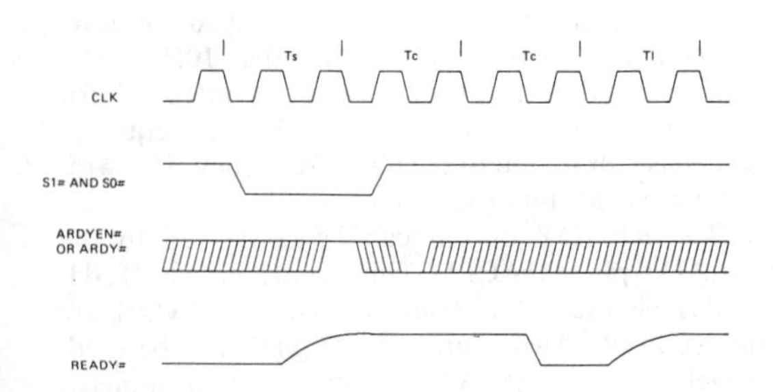

Figure 7.8 *The ARDY asynchronous input timing diagram. SRDYEN must be held high, and for another READY both ARDYEN and ARDY must be taken LOW.*

The 82C284 can also be used for providing signals for the 80386DX and 80386SX processors, but for some specialized circuits, particularly when clock frequencies higher than 25 MHz are used, other oscillators (even discrete circuits) may have to be used. See also the description of the 82230/82231 chip set later in this section.

Bus controllers

The bus controller for the 8088/8086 series is the 82C88, whose pin-out for a 20-pin DIL chip is illustrated in Figure 7.9. This is a replacement for the older bipolar 8288 chip which is still used in con-

The # sign means that the pin is active LOW

Pin	Signal	Pin	Signal
1	IOB	20	V_{CC}
2	CLK	19	S0#
3	S1#	18	S2#
4	DT/R#	17	MCE/PDEN#
5	ALE	16	DEN
6	AEN#	15	CEN
7	MRDC#	14	INTA#
8	AMWC#	13	IORC#
9	MWTC#	12	AIOWC#
10	GND	11	IOWC#

Figure 7.9 *The pin-out for the 82C88 bus controller, which uses a 20-pin DIL package.*

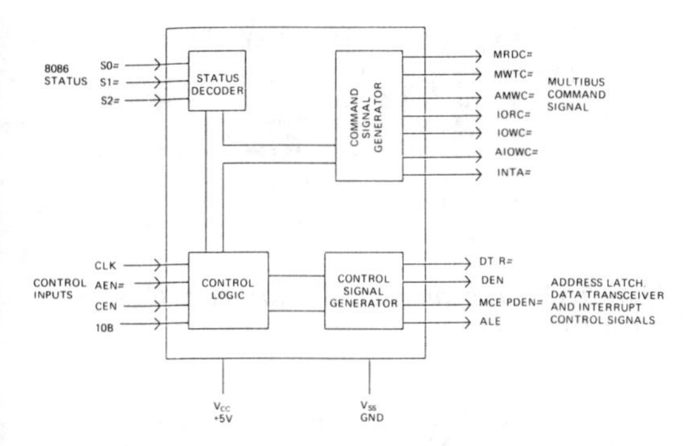

Figure 7.10 *A block diagram for the 82C88 bus controller.*

trollers. The chip is used for maximum mode systems, and uses the status and control signals from the 8088/8086 CPU to generate the standard bus signals of MRDC, MWTC, AMWC, IORC, IOWC, AIOWC and INTA (all active LOW). The latch, transceiver and interrupt control signals DT/R, DEN, MCE/PDEN and ALE are also generated. The standard version of the chip is the 82C88–2 which can use a clock frequency of 8 MHz. The 82C288 uses static circuitry, so that there is no minimum clock frequency stipulated. The clock can even be stopped without loss of signal status.

Figure 7.10 shows the block diagram, illustrating how the chip will be used in the 8088/8086 maximum mode system. The way in which the outputs of the 82C88 are determined from the CPU status signals is shown in Figure 7.11.

The bus controller for the 80286 CPUs is the 80288

8088/8086 Processor				
Inputs			Processor state	82C88 Outputs
S2#	S1#	S0#		
0	0	0	Interrupt acknowledge	INTA#
0	0	1	Read I/O port	IORC#
0	1	0	Write I/O port	IOWC#,AIOWC#
0	1	1	HALT	None
1	0	0	Code read from memory	MRDC#
1	0	1	Read memory	MRDC#
1	1	0	Write memory	MWTC#,AMWC#
1	1	1	Passive	None

Figure 7.11 *How the outputs of the 82C84 are determined from the status signals.*

The # sign means that the pin is active LOW

Pin	Signal	Pin	Signal
1	READY#	11	IOWC#
2	CLK	12	IORC#
3	S1#	13	INTA#
4	MCE	14	CENL
5	ALE	15	CEN/AEN#
6	MB	16	DEN
7	CMDLY	17	DT/R#
8	MRDC#	18	M/IO#
9	MWTC#	19	S0#
10	GND	20	V_{CC}

Figure 7.12 *The pin allocation for the 80C288 as a 20-pin DIL package.*

or 80C288, and since the two are completely compatible, the later CHMOS version will be dealt with here. The chip is available as 8, 10 or 12 MHz versions, distinguished by the suffix to the chip number. The pin arrangement is either as a 20-pin DIL or the 20-pin PLCC package whose pin assignment is illustrated in Figure 7.12. The block diagram, Figure 7.13, shows that the inputs are the status group (S0, S1 and M/IO) and the control group of CEN/AEN, CENL, CMDLY, READY and MB. The outputs are the bus set that are also the outputs of the 82C88.

The 82C288 MB input is used to control two modes of bus operation. When the MB input is strapped HIGH, the bus timing is to the MULTIBUS standard, meeting IEEE-796 specifications, and requiring at least one wait state in the bus cycle. With the MB input LOW, there are no delays and no wait states are necessary for correct bus operation.

The actions of the S0, S1 and M/IO inputs in terms of activating the bus output signals are illustrated in Figure 7.14. The status signals always change in the first phase of the local bus clock, so that they serve

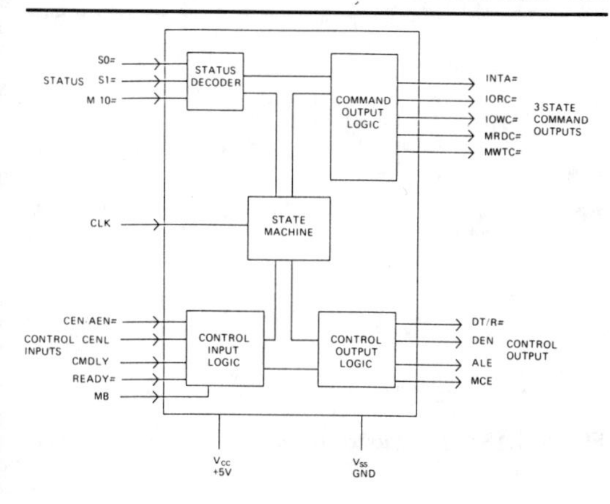

Figure 7.13 *A block diagram for the 80C288.*

M/IO#	S1#	S0#	Type of Bus cycle
0	0	0	Interrupt acknowledge
0	0	1	I/O Read
0	1	0	I/O Write
0	1	1	None; Idle
1	0	0	HALT or Shutdown
1	0	1	Memory Read
1	1	0	Memory Write
1	1	1	None; Idle

Figure 7.14 *The actions of the S0, S1 and M/IO#* signals on the bus controller.

also to correct the phase of actions controlled by the bus master chip.

The controller permits three bus states, Ti, Ts and Tc, whose relationships are illustrated in Figure 7.15. Each state will be 2 CLK cycles in duration, corresponding to two phases of the CPU clock. The Ti state (idle) occurs when the local bus is inactive, and this can be continued indefinitely. The bus controller maintains an idle state when the bus control is being passed over from one master to another.

The start of a bus use cycle is signalled by either S0 or S1 input going LOW, starting a Ts state which will run for the two CLK cycles in which the S0 or S1 signals are active. These inputs are sampled at each falling edge of the CLK pulse and when the result of this sample returns LOW for either input, the next CLK cycle is taken as being the second phase of the internal CPU clock.

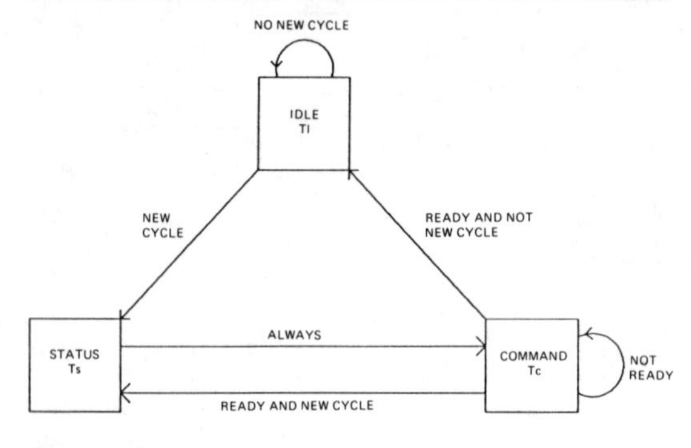

Figure 7.15 *The bus states of the 80C288 and their relationships.*

The Ts state is always followed by the Tc state, which may be repeated to provide for waits. The READY signal is sampled at the end of each Tc state and if this signal is LOW, the bus cycle is terminated. When this happens, the Idle state will follow if no new cycle is waiting, or the Ts state will follow if a new cycle is pending.

Where a system uses more than one bus (apart from the local bus), one bus controller will be used for each bus, with only one controller active at any particular time. If a MULTIBUS system is in use, then each bus may be controlled by more than one CPU, with only one CPU in control of the bus at any given time. For systems in which more than one 82C288 is in use, the CENL and AEN control inputs are used as enables. The CENL input (active HIGH) is used as a chip enable to permit a bus controller to be in command of a bus cycle. The AEN input (active LOW) permits the controller to drive outputs so that when this input is put high another bus controller can be used to provide the bus output signals.

The 82C288 can be used along with the 80386DX and 80386SX, and another possibility is the use of the integrated 82230/82231 chip set described later in this section.

DMA controllers

The DMA controller allows for the transfer of data, usually involving memory, without the intervention

```
Pin configuration - the # sign indicates active LOW.

 1. IOR#        40. A7
 2. IOW#        39. A6
 3. MEMR#       38. A5
 4. MEMW#       37. A4
 5. PIN 5       36. EOP#
 6. READY       35. A3
 7. HLDA        34. A2
 8. ADSTB       33. A1
 9. AEN         32. A0
10. HRQ         31. Vcc
11. CS#         30. DB0
12. CLK         29. DB1
13. RESET       28. DB2
14. DACK2       27. DB3
15. DACK3       26. DB4
16. DREQ3       25. DACK0
17. DREQ2       24. DACK1
18. DREQ1       23. DB5
19. DREQ0       22. DB6
20. GND         21. DB7
```

Figure 7.16 *Pin assignment for the 8237 DMA controller, which uses a 20-pin DIL package.*

of the CPU. The transfers that can be carried out are memory to memory, memory to I/O and I/O to memory. DMA controllers are programmable devices, and since any discussion of software is ruled out for lack of space in this section, this description will concentrate on the hardware aspects of the various DMA controllers in the Intel chip set.

The 8237A and 82C37A-5 are the main versions of the DMA controller that was developed for the 8088/8086 systems. Since the CHMOS version is compatible with the earlier NMOS 8237A-5 type, but is fully static, with much lower power consumption, this version only will be described.

The chip takes the form of a 40-pin DIL package, whose pin-out is shown in Figure 7.16. The chip contains six 16-bit registers, four 8-bit registers, one 6-bit register and two 4-bit registers, all used to program the chip for the desired transfers. The V_{cc} input on pin 31 requires a +5 V supply, with the V_{ss} (pin 20) returned to earth/ground. Pin 5 (known as PIN5) should always be strapped to V_{cc}, even though it has an internal pull-up resistor.

The CLK input controls the timing of the internal actions and also controls the rate of data transfers. The maximum clock rate for the 82C37A-5 is 5 MHz, so that in some systems, the clock output from the CLK output of the clock chip may be too fast and the PCLK output needs to be used. The other inputs are as follows:

CS, active LOW, pin 11, is a chip select which will activate the 82C37A as a port during the idle bus cycle, allowing the CPU to communicate on the data bus. RESET and READY correspond to the identically-labelled pins on the CPU and are used in the same way. The READY input is used in the usual way to make the DMA controller perform wait states to allow for the use of slow memory or I/O devices. The use of RESET clears all registers, and puts the controller into its idle cycle. The HLDA input is a hold acknowledge from the main CPU which is active HIGH to indicate that the system buses can be used by the DMA controller.

The DMA request lines are DREQ0 to DREQ3, in ascending order of priority. Each of these inputs is active HIGH after a RESET (the polarity of activity can be programmed) and a request is generated by a peripheral making the voltage on one of these pins HIGH. This will be acknowledged by the appropriate DACK output (see later). The DREQ request signal voltage must be maintained until the DREQ is obtained. If a low-priority request conflicts with one of higher priority, the higher-priority request will be serviced first.

The outputs of the DMA controller are as follows:

DACK0 to DACK3 are DMA request acknowledge outputs which are used as a handshake between the DMA controller and its peripherals, along with the DREQ inputs. Following a RESET these lines are active LOW, but their polarity can be reprogrammed in the course of using the DMA controller.

The A4–A7 lines are the upper four address lines – remember that I/O requires only 8-line addresses, and for memory control the upper 8-bits of a complete memory address **within a segment** are latched. The CPU must control the segment address bits on A16–A19 at all times. The A4–A7 outputs are enabled only during a DMA operation and are floated at all other times. The AEN address enable, along with the ADSTB (active HIGH) strobe is used to latch the upper eight address bits to the A0–A15 portion of the address bus. These bits are latched from the data bus, on to which they have been placed by reading the address register of the DMA controller.

The HRQ output from the DMA controller, active HIGH, is used to request use of the buses, and is issued after an unmasked DMA request has been received. The HLDA will be issued from the CPU at least one clock cycle later (if the CPU is not currently executing an instruction), but the delay will be greater if the CPU is active.

The other two outputs are active LOW. The MEMR output goes LOW to provide access to data from memory that is to be read during a DMA read or memory-to-memory transfer. The MEMW signal goes LOW to write data to a selected address during a DMA write or a memory-to-memory transfer cycle. Both of these outputs are floated when not in use.

The remaining lines are used both for inputs and outputs. The A0–A3 address lines are used in the idle cycle as inputs so that the CPU can gain access to the registers by using these 4 bits as a register address. In active cycles, these are outputs which provide the lower 4 bits of the output address, I/O or memory.

The data bus DB0–DB7 lines are also bidirectional, and the use of these lines depends on the way in which the chip is being used. The data lines will transmit the contents of DMA registers to the CPU during the I/O read, and will be used to program the registers during the time when the CPU is writing to the DMA controller. In a DMA cycle, the most significant byte of the address will be output along these lines, and strobed into a latch to be placed on to the address lines. When memory-to-memory transfers are being used, data read from memory will be input along the data lines in one half of the transfer and output to the new memory location in the second half.

The remaining bidirectional lines are used for control signals. The IOR signal is active LOW, and in the idle cycle is an input which is used by the CPU to read the control registers of the 82C37A. In active cycles, this signal is an output which is used to enable a read from a peripheral during a DMA write cycle (write to memory from I/O). The corresponding IOW signal is also active LOW. On idle cycles it is used by the CPU to control the loading of data into the DMA controller. On active cycles, it is an output control signal which allows data to be loaded to a peripheral during a DMA read (read memory to I/O).

The remaining line is EOP, also active LOW. An external signal can terminate a DMA action by pulling this signal input voltage LOW. The alternative is that the DMA process is completed when the terminal count for any channel is reached, causing the EOP pin voltage to go LOW as a signal to external devices. If this pin is not used (as is often the case) it should be tied to V_{cc}.

The 82C37A operates in two types of cycles, idle cycles and active cycles, with several states possible in each cycle; up to 6 states are possible, labelled as S0–S4 and Sw. Each state occupies one complete clock cycle. The S1 state is inactive, and in the idle cycle of the DMA controller S1 states are repeated while no valid DMA requests are pending. During the idle cycle, the DREQ lines will be tested on each clock cycle to find if any channel is requesting DMA. The CS input will also be sampled to find if the CPU is attempting to read from or write to the registers of the DMA controller.

In the S1 state, the DMA controller can be programmed by the CPU (CS and HLDA both LOW), using address lines A0–A3 as inputs to select the internal registers of the 82C37A. The IOR and IOW lines are used for selecting and timing reading and writing respectively. Since 6 of the internal registers are 16-bit, and only 8 data lines are available, an internal flip-flop switches between upper and lower byte of the 16-bit registers. This flip-flop can be reset by software or by a RESET. In this *program condition*, a few software commands can be executed which do not make use of the data bus. Figure 7.17 shows the software codes which can be used to write and read the internal registers.

A3	A2	A1	A0	IOR#	IOW#	Operation
1	0	0	0	0	1	Read status register
1	0	0	0	1	0	Write command register
1	0	0	1	0	1	ILLEGAL
1	0	0	1	1	0	Write request register
1	0	1	0	0	1	ILLEGAL
1	0	1	0	1	0	Write single mask register bit
1	0	1	1	0	1	ILLEGAL
1	0	1	1	1	0	Write mode register
1	1	0	0	0	1	ILLEGAL
1	1	0	0	1	0	Clear byte pointer flip-flop
1	1	0	1	0	1	Read temporary register
1	1	0	1	1	0	Master clear
1	1	1	0	0	1	ILLEGAL
1	1	1	0	1	0	Clear mask register
1	1	1	1	0	1	ILLEGAL
1	1	1	1	1	0	Write all mask register bits

Figure 7.17 *Programming the DMA controller registers.*

When the DMA controller requests a bus hold, the S0 state is entered, but in the time before the HLDA is active, the DMA controller can still be programmed by the CPU. A DMA transfer (I/O to or from memory) makes use of States S2 to S4, returning to S1. If wait states are needed, Sw states can be inserted between S2 and S3 or between S3 and S4 by using the READY line. Transfers between I/O and memory do not involve any transfer of data into or out of the 82C27A.

In a memory-to-memory transfer there are two main sections, the read from memory and the write to memory, and each will take at least four states. For such cycles, the states are labelled S11–S14 for reading and S21–S24 for writing. In all types of transfers, bits A8–A15 are latched in S1, and the latching will often not require to be changed if these bits do not change during a set of byte transfers. This means that the S1 states need be executed only when a change in the A8–A15 bits is needed, which can be once in 256 transfers, so saving 255 clock cycles in 256 byte transfers.

The registers of the 82C37A are illustrated in Figure 7.18 with a brief description of the operation of each. DMA action can be carried out in any of four modes. In single transfer mode, the controller will transfer one word only, decrementing the count register and decrementing or incrementing the address on each transfer. Another option is block mode, in which transfers will be made until the word count register contains FFFFH or the transfer is stopped by receiving the EOP signal.

In demand transfer mode, transfers can be continued until an external I/O device has no more data to transfer, and the CPU can regain control until more data is ready to be transferred. In this mode, the EOP signal is the usual way of stopping a transfer, either by hardware or by software. In cascade mode, more than one 82C37A can be used in a chain, with the main DMA controller not providing any address or control signals, only exercising priority control.

Three different types of transfer may be made in any of the three transfer modes (cascade is not a transfer mode, but a way of controlling other DMA controllers). These types are read, write and verify, and the names read and write refer to the reading or

All Registers are 8-bit:

7	6	5	4	3	2	1	0

Command Register bits:

Bit	Action for 0	Action for 1	Note
0	Mem-Mem disable	Mem-Mem enable	
1	Ch.0 address hold disable	Ch.0 address hold enable	1
2	Controller enable	Controller disable	
3	Normal timing	Compressed timing	2
4	Fixed priority	Rotating priority	
5	Late write selection	Extended write selection	3
6	DREQ sense active HIGH	DREQ sense active LOW	
7	DACK sense active LOW	DACK sense active HIGH	

1. 0 or 1 if bit0=0
2. 0 or 1 if bit0=1
3. 0 or 1 is bit3=1

Mode Register bits:

Bits	Pattern	Action
10	00	Channel 0 select
	01	Channel 1 select
	10	Channel 2 select
	11	Channel 3 select
32	00	Verify transfer
	01	Write transfer
	10	Read transfer
	11	ILLEGAL
	XX	if bits 6,7 = 11
4	0	Autoinitialization disable
	1	Autoinitialization enable
5	0	Address increment select
	1	Address decrement select
76	00	Demand mode select
	01	Single mode select
	10	Block mode select
	11	Cascade mode select

Request Register

Bits	Pattern	Action
10	00	Select Channel 0it
	01	Select Channel 1
	10	Select Channel 2
	11	Select Channel 3
2	0	Clear request bit
	1	Set request bit

Mask Register:

Bits	Pattern	Action
10	00	Channel 0 mask bit
	01	Channel 1 mask bit
	10	Channel 2 mask bit
	11	Channel 3 mask bit
2	0	Clear mask bit
	1	Set mask bit

In this method of use, bits 3 - 7 are ignored
NOTE: As an alternative, bits 0 - 3 can be used as individual set or clear bits for channels 0 - 3.

Status Register:

Bit	Action for 1
0	Channel 0 has reached terminal count
1	Channel 1 has reached terminal count
2	Channel 2 has reached terminal count
3	Channel 3 has reached terminal count
4	Channel 0 request
5	Channel 1 request
6	Channel 2 request
7	Channel 3 request

Figure 7.18 *The register map of the 82C37A.*

Figure 7.18 (*continued*)

Register Codes:

Register	Operation	Signals						
		CS#	IOR#	IOW#	A3	A2	A1	A0
Command	Write	0	1	0	1	0	0	0
Mode	Write	0	1	0	1	0	1	1
Request	Write	0	1	0	1	0	0	1
Mask	Set/Reset	0	1	0	1	0	1	0
Mask	Write	0	1	0	1	1	1	1
Temporary	Read	0	0	1	1	1	0	1
Status	Read	0	0	1	1	1	0	0

writing of **memory** so as to write to or read from an I/O device.

Memory-to-memory transfer is selected by programming the command register to use channels 0 and 1, reading from one memory address or block of addresses to another set of addresses. In the auto-initialize transfers, each EOP will cause the original values of current address and current word count to be restored in the registers. The priority of channels may be fixed, using the sequence of 0–3 as the order of decreasing priority. Another option is rotating priority in which the last channel to be serviced becomes the lowest priority, with the others in the sequence rotating their positions accordingly.

Compressed transfers can be used to decrease the

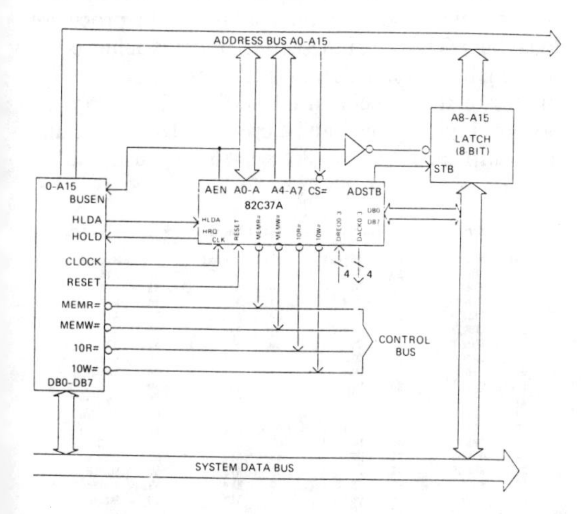

Figure 7.19 *A typical circuit using the 82C37A controller, showing an interface with the older 8080 8-bit CPU.*

time for a transfer as the system permits, by removing the S3 state and so making the read pulse time equal to the write pulse time.

Figure 7.19 shows the DMA controller in a typical circuit, assuming the use of a CPU with demultiplexed address and data lines (in this case the 8-bit 8080 or 8085 CPU).

The 82258 ADMA Co-processor

The 82258 can be used in 8088/8086, 80188/80186, or 80286 circuits as a DMA device which can be used as a co-processor to the 80286, and also for the 80386. The 6 MHz and 8 MHz options allow for fast operation, with a maximum transfer rate of 8 Mb/s at the faster clock rate. The full 80286 memory of 16M can be addressed, with up to four DMA channels. One channel can be programmed to handle up to 32 subchannels at slower rates.

The 82258 is available in 68-pin packages as a leadless chip carrier package or as a pin-grid array; the pin assignments are shown in Figure 7.20. There are four operational modes, for 80286 use, for 80188/80186 use, for 8088/8086 maximum mode use, or for remote use (on a bus that is not the processor local bus). The use of these four modes is achieved by making the pin signals match the bus structure of the processor that is being used, so that, for example, a non-multiplexed bus can be used in 80286 mode and a multiplexed bus in 8088/8086 or 80188/80186 mode.

The hashmark (#) is used to identify pins whose action is active LOW.

Pin	Signal	Pin	Signal	Pin	Signal	Pin	Signal
1	BHE#	18	D15	35	A0	52	A16
2	RD#	19	D7	36	A1	53	A17
3	WR#	20	D14	37	A2	54	A18
4	DREQ3	21	D6	38	A3	55	A19
5	DREQ2	22	D13	39	A4	56	A20
6	DREQ1	23	D5	40	A5	57	A21
7	DREQ0	24	D12	41	A6	58	A22
8	CS#	25	D4	42	A7	59	A23
9	V_{SS}	26	V_{CC}	43	V_{SS}	60	V_{CC}
10	READY#	27	D11	44	A8	61	DACK0#
11	S1#	28	D3	45	A9	62	DACK1#
12	CLK	29	D10	46	A10	63	DACK2#
13	S0#	30	D2	47	A11	64	DACK3#
14	M/IO#	31	D9	48	A12	65	EOD0#
15	RESET	32	D1	49	A13	66	EOD1#
16	HOLD	33	D8	50	A14	67	EOD2#
17	HLDA	34	D0	51	A15	68	EOD3#

Figure 7.20 *The pin assignment for the 82258 ADMA co-processor. The pin–grid is a 68-pin type as illustrated in Figure 1.5(b).*

Pins common to all modes:

Pin Name	I or O	Action(s)
BHE#	I/O	Enables transfers on upper byte of bus
DREQ3#	I	DMA request, channel 3
DREQ2#	I	DMA request, channel 2
DREQ1#	I	DMA request, channel 1
DREQ0#	I	DMA request, channel 0
S1#,S0#	I/O	Status signals
RESET	I	Reset to initial state
DACK3#	O	DMA request acknowledge, channel 3
DACK2#	O	DMA request acknowledge, channel 2
DACK1#	O	DMA request acknowledge, channel 1
DACK0#	O	DMA request acknowledge, channel 0
EOD3#	I/O	End of DMA, channel 3
EOD2#	I/O	End of DMA, channel 2
EOD1#	I/O	End of DMA, channel 1
EOD0#	I/O	End of DMA, channel 0

Pin uses in 80286 mode:

Pin Name	I or O	Action(s)	Pin(s)
RD#	I	With CS# reads 82258 register	2
WR#	I	With CS# writes 82258 register	3
CS#	I	Chip select used to access registers	8
READY#	I	Terminates a bus cycle	10
CLK	I	System clock input	12
M/IO#	O	Selects memory or I/O port	14
HOLD	O	Requests control of local bus	16
HLDA	I	Signals that 82258 can take buses	17
D0-D15	I/O	Bidirectional local data bus	18-25,27-34
A0 - A7	I/O	Lower address lines for DMA, registers	35-42
A8-A23	O	Remaining bus lines (A21 needs pull-up)	44-59

Pins used for 80186 mode (differences as compared to 80286 mode):

Pin Name	I or O	Action(s)	Pin(s)
RD#	I/O	Read in 80186 mode	2
WR#	I/O	Write in 80186 mode	3
ALE	O	Address latch strobe	58
DEN#	O	Data enable for transceivers	56
DT/R#	O	Transceiver direction (needs pull-up)	57
SREADY	O	Synchronous READY signal	10
CLK	I	System clock input	12
S2#	O	Status, with S1 and S0	14
AD0-AD15	I/O	Multiplexed address and data	18-24,27-34
A0 - A7	I/O	Demultiplexed address lines	35-42
A8-A15	O	Demultiplexed address lines	44-51
A16/S3	O	Multiplexed address/status line	52
A17/S4	O	Multiplexed address/status line	53
A18/S5	O	Multiplexed address/status line	54
A19/S6	O	Multiplexed address/status line	55
AREADY	I	Asynchronous READY	59

Pin used in 8086 mode (differences from 80186 mode):

Pin Name	I or O	Action(s)	Pin(s)
RQ#/GT#	I/O	Request/grant for system bus	16
HLDA	I	Tie HIGH - no meaning for 8086	17
CS#	I	Chip select for registers or bus access	8
BREL	O	Bus release, 8225 has released buses	14

Figure 7.21 *Pin signals summary for the 82258.*

The pin signal description, excluding power supplies, is summarized in the tables of Figure 7.21, showing first the pin signals that are common to all modes, followed by the specialized uses of pins in the four modes. The format of many of these signals follows that of the processor bus to which the

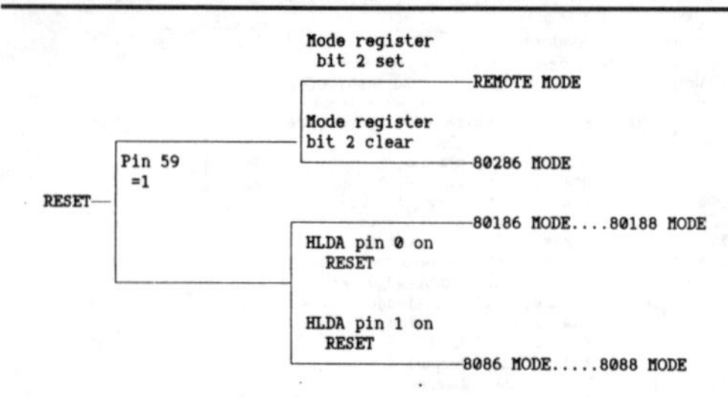

Figure 7.22 *How the register contents affect the mode of operation of the 82258.*

controller will be connected, and others will be already familiar from the description of the 82C37A. The different modes of use are determined by the A23 and HLDA pins, plus the RM bit in the mode register of the chip – Figure 7.22 shows how the various modes can be determined. The 286 mode can also be used to configure the chip for use with the 80386 processors.

Programming the registers of the 82258 is done by using what is described as the 'slave interface', consisting of the S0, S1 status lines, RD, WR control lines, A0–A7 register address lines, D0–D15 data lines (286) or the AD0–AD15 address/data lines (for 8088/8086 or 80188/80186 modes). This set of interfacing lines permits either synchronous or asynchronous transfers, using S0 and S1 for synchronous transfers and the RD, WR lines for asynchronous. The register address within the 82258 is obtained using the A0–A7 (or AD0–AD7) lines. In remote mode, only asynchronous transfer is possible because the 82258 must release the local bus to allow the CPU to gain access to its registers.

The register map for the 82258 is illustrated in Figure 7.23. There are five general registers, two of 16-bits and three of 8-bits each, which apply to the overall control of the chip. Each of the four channels uses a set of 11 registers, of which seven are 24-bit, three are 16-bit and one is 8-bit. For multiplex channel operation, one of the 24-bit registers for channel 3 is used to point to the position of the multiplexer table in memory, and three other 8-bit registers are used to handle interrupt vector, last vector and subchannel numbers.

The CPU can read the registers of the 82258 also in

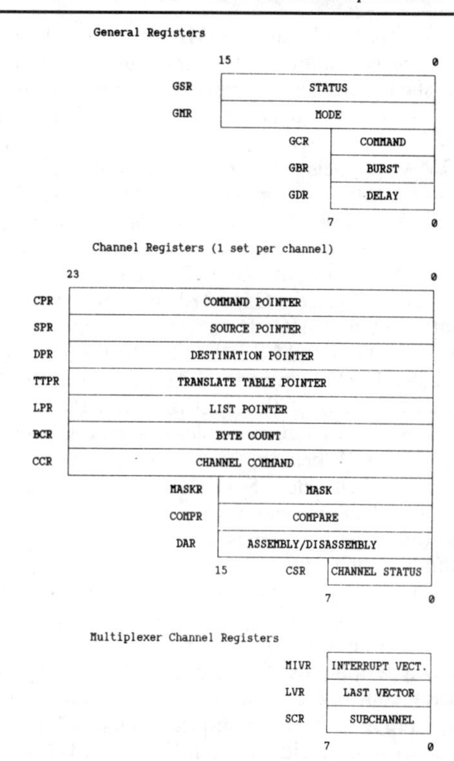

Figure 7.23 *The register map of the 82258.*

two different ways. The hardware method uses one or more of the EOD lines (active LOW) as interrupt request lines to the CPU. Following such an interrupt, correctly vectored, the CPU can read the status registers of the 82258 and service the interrupt. The alternative uses the system called 'control space communication', in which the 82258 writes the contents of a channel status register into a channel command block at the end of a DMA transfer.

The communication with peripherals is carried out using the DREQ, DACK and EOD signals. The DREQ and DACK pair (any of four) will carry out the handshaking with the peripheral, and data can be transferred for as long as the request is active. The end of DMA is signalled by the EOD, active LOW. The EOD pin can be used as an input for a signal that will asynchronously end a DMA transfer. The pin can also be used as an output to interrupt the CPU to signal such events as an aborted transfer, end of a block or request for next block. The EOD for channel 2 can be used as a single interrupt for all of the DMA channels if this method is adopted.

Bus arbitration can make use of HOLD/HLDA for 80286, 80188/80186, or for minimum mode 8088/8086 systems. For use with 8088/8086 maximum mode systems, the RQ/GT signals can be used, and for remote use the CS/BREL system is used. In this latter method, HOLD and HLDA will be used on the system bus and CS/BREL on the local bus (of the 82258). The CPU puts the CS signal active LOW to request the use of the local bus, and the 82258 releases the bus when possible and places BREL active HIGH. When the CPU has completed its transfers it then sets CS HIGH again, following which the 82258 will put BREL LOW.

The four DMA channels are independently programmable, each with its own register set. Channel 3 can be used in the same way as the others, or alternatively as a multiplexed channel to allow communication with a large number of (slow) port-addressed peripherals. This is done using the 8259A interrupt controllers to arbitrate and control the channel requests in the multiplexed channel, using the type of scheme illustrated in Figure 7.24. Multiplexing is controlled by using a table set up in external memory (the multiplexer table, MT) to store command pointers and the 8259A mask register locations for each device.

The DMA transfers are of two basic types, single-cycle or two-cycle transfers. In this sense, *cycle* does not refer to a bus cycle but a cycle of 82258 activity. In a single-cycle transfer, bytes or words are read from a source and written directly to a destination without being stored at any point in the registers of the 82258. The fastest rate of transfer that can be achieved by this method is 8 Mb/s on the 8 MHz 82258 connected in an 80286 system. For the 80186 (or similar) 8 MHz system, the maximum speed is 4 Mb/s.

Single-cycle transfer cannot be used along with multiplexing on channel 3, though if channel 3 is not being used as a multiplex channel it will support single-cycle transfer like the other channels. The

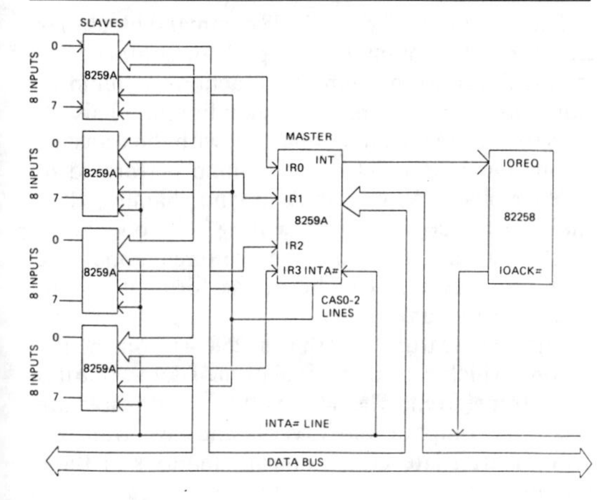

Figure 7.24 *Using the multiplexer channel with interrupts from a set of 8259A controllers.*

single cycle starts with DACK being pulled LOW to acknowledge that the cycle will start. At the same time, the pointer to the destination location (memory or port) is issued and the 82258 executes a complete bus cycle. The appropriate read and write command signals have to be available from the devices that are being used as source and destination.

The two-cycle transfer involves reading a byte or word into the 82258 registers and then writing it out again. This can be particularly convenient if a word has to be read and 2 bytes written or 2 bytes read and a word written, allowing 8-bit and 16-bit systems to be linked. This scheme is used for the multiplexed channels, and also for memory to memory transfers. Another application is to 'no-address' transfers, meaning the transfer of a byte or word to or from the 82258 data assembly register.

The 82258 exercises its control by way of a command block that has to be set up in the external memory for each channel. When the CPU initiates the channel start sequence, the 82258 will read the channel control block parameters from memory into the registers for that channel within the chip. All registers that can be programmed can be altered by way of the CPU, but the initialization from a command block is often the only access that is required.

The commands for the 82258 command blocks are either transfer commands (Type 1) or control commands (Type 2), meaning such actions as jump or stop, either conditional or unconditional. This allows for commands to be chained, with the transfers of one block completed and a jump performed to another block. As well as command chaining, data chaining can be used to allow the 82258 to work on blocks or data that may be scattered about the memory. Chaining can be in the form of a continuous list or as a linked list.

Another feature of the 82258 is 'on-the-fly' actions, which permit manipulation of data while it is being transferred. The actions that are possible are mask and compare, verify, verify and save, translate (using a translate table), or combinations of these actions.

Interrupt controllers

The 8-bit and 16-bit chips from 8080 to 8086 can make use of the 82C59A-2 interrupt controller, a development of the 8259A with which it is compatible. Like the earlier version, the CHMOS chip is entirely static and needs no clock input. It allows up to 8 vectored priority interrupts to be managed, and can be used in cascade without additional circuity to handle up to 64 interrupt sources. For any interrupt, the controller will determine which of a set of simultaneous interrupt requests has greatest priority and will issue an interrupt, passing also the vector address that is required to service the interrupt. Like many other support chips, the interrupt controller is programmable, making use of internal registers that are accessible by the CPU.

The block diagram is illustrated in Figure 7.25, showing how an internal bus is used to connect registers, and how the inputs and outputs are organized. The pinout is for a 28-pin DIL package (Figure 7.26).

Pins 14 and 28 are used for V_{ss} (ground/earth) and V_{cc} (+5 V) respectively. Four inputs are used for selection, being the chip select (CS, pin 1), WR write enable (pin 2), RD read enable (pin 3) and A0 address input. The CS, WR and RD inputs are all active LOW, A0 is active HIGH. The CS input must

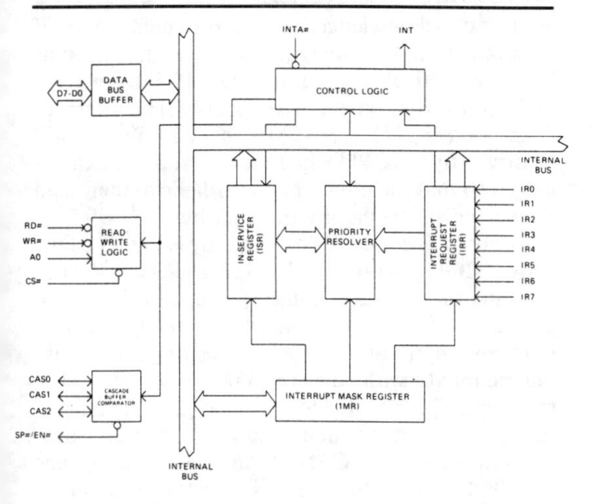

Figure 7.25 *Block diagram of the 82C59A interrupt controller.*

The # sign means that the pin is active LOW

Pin	Signal	Pin	Signal
1	CS#	28	V_{CC}
2	WR#	27	A0
3	RD#	26	INTA#
4	D7	25	IR7
5	D6	24	IR6
6	D5	23	IR5
7	D4	22	IR4
8	D3	21	IR3
9	D2	20	IR2
10	D1	19	IR1
11	D0	18	IR0
12	CAS0	17	INT
13	CAS1	16	SP#/EN#
14	GND	15	CAS2

Figure 7.26 *The pin assignment for the 28-pin DIL package.*

be LOW in order to allow the WR and RD signals to be recognized, and the A0 input line is used to enable CPU reading and writing of the chip. This A0 pin will be connected to the A0 address line of 8-bit CPUs or to the A1 line of the 8088/8086 CPUs. The WR input permits the CPU to write command words into the interrupt controller, and the RD signal enables the CPU to read status information from the interrupt controller.

Interrupt requests are placed on the IR0–IR7 lines, all active HIGH – the level must stay HIGH at least until acknowledged, and can remain HIGH if required until the interrupt is serviced. The pins 18–25 respectively are maintained HIGH by internal pull-up resistors unless driven LOW. The interrupt acknowledge (INTA, pin 26) is active LOW, and will be driven by the CPU when the interrupt is acknowledged so that the interrupt controller can then send the vector address along the data bus.

The only line used for output only is the INT line, active HIGH, on pin 17. This signal will go HIGH to indicate that a valid interrupt request has been granted, and it will be connected to the (maskable) interrupt input of the CPU. The other lines are bidirectional, with the D0–D7 lines (pins 11–4 respectively) connected to the corresponding lines of the system data bus and used for vector and programming data. The CAS0–CAS2 cascade lines (pins 12, 13, 15) are used in a bus structure to connect multiple 82C59 chips so that one master 82C59 can control a number of slave chips. The SP/EN pin (pin 16) can be used as an input to force the chip to be a master (SP = 1) or slave (SP = 0). The alternative use of this pin in buffered mode (see later) is to control transceivers.

The registers of the 82C59A consist of the in-service register (ISR), the interrupt request register (IRR) and the interrupt mask register (IMR), all of eight bits each. These registers are written by using command words from the CPU (rather than by addressing them directly). The IRR shows the levels of requested interrupts, with a bit set for each interrupt and the position of the bit denoting priority level. The ISR shows which interrupts are being serviced, and when a bit is set in the ISR to indicate servicing, the corresponding bit will be reset in the IRR. The IMR has its various bits set to indicate masking of various levels of interrupt.

The CPU issues two types of command words, the initialization command words (ICW) and the operation command words (OCW). The ICW bytes (typically 2 to 4 bytes) are used to prepare the chip for use, with the initialization bytes being timed by the WR pulses. The command words select the mode of operation from fully nested, rotating priority, special mask mode or polled mode.

The fully-nested mode is the default following initialization. The interrupt requests have the priority order of 0 (highest) to 7 (lowest). When an interrupt is acknowledged, the highest priority of request is selected and its vector address placed on the bus. At the same time, the corresponding bit in the ISR is set (and the IRR bit is reset). This has the effect of disabling any interrupts of the same or lower priority. Interrupts of a higher priority will be serviced if the software that serviced the first interrupt has re-enabled the interrupt system of the CPU.

The rotating priority mode allows for all the interrupts being of about equal priority, so that the most recent interrupt is automatically put as lowest priority and all others are shuffled around one place. The special masking mode allows a mask bit to be set, allowing no interrupts at that level, but enabling interrupts from all other levels, lower as well as higher. In poll mode, the INT at the CPU must be masked so that polling is used rather than interrupts to check each input in turn for data.

When 82C59A controllers are operated in master–slave fashion, the CAS0–CAS2 pins of the chip that has been designated as master (SP/EN HIGH) are connected to the corresponding pins of the chips that have been designated as slaves (SP/EN LOW).

The 82C59 controllers are also used in 80286 systems and can be used in 80386 systems also.

Support for the 80286 and 80386 family

By the time the 80386 was developed, the support chips for the 8088/8086 and the 80286 chips already existed, and the chip count for a computer using these chips was fairly large. This led to specific support chips being manufactured for the 80286 and 80386 to reduce the chip count by compressing the functions of several formerly separate chips into single units. Since the actions that are carried out by these units are those that have been described for the individual support chips, there is no point in repeating these descriptions, and the most recent support chips are therefore described in terms of the tasks that they can carry out.

The support chips for the 80386 fall into three main groups, the chip set for the AT type of bus, the chip set for MCA buses, and the very recent EISA bus chip set. These are all of primary interest to designers of computer systems, and the differences are too specialized to deal with here. The chip set for the AT bus, the most popular form of bus structure, will be dealt with here, and the chips for the MCA form of buses noted.

The AT bus chips are the 82230 and 82231, both of which are physically large LSI chips. The 82230 is illustrated in block diagram form in Figure 7.27, and this shows also the huge range of functions that are covered by this chip, which uses an 84-pin PLCC package. The block diagram for the 82231, which also uses the 84-pin PLCC package, is shown in Figure 7.28 – note that the 82231 contains the equivalent of the 8284 clock generator, but the 82230 contains the equivalent of the 82284 clock, which can be used to synchronize the clock of the 82231.

The two chips depend on each other for a few actions, so that 14 pins on each are used for intercommunication. For most purposes, however, the chips are independent, with the 82230 generating most of the timing and control signals and the 82231 controlling the X address bus (see below) for DMA and refresh. Co-processor interfacing logic is contained in the 82230.

The AT bus system uses five forms of address bus, each independently buffered. These address buses are designated local, system, memory, X and L. There are also four data buses, local, system, memory and X-data. The purpose of using these buses is specific to the IBM PC/AT format of computer, and this style of bus system has become a standard for a huge number of small computers using either the 80286, 80386DX or 80386SX CPUs. Later systems using the MCA or EISA buses are noted briefly later in this section.

Examining the address buses first, the local address bus uses the 24 address pins of the 80286 or the 80386SX. The A0 line is directly connected to the 82230 but all other lines are buffered and latched. The system address bus is the main computer address bus, consisting of 20 lines, one of which, SA0, is obtained from the 82230, the others being latched from the local bus by the ALE signal from the 82230. The HLDA signal is used as the output enable.

The memory address bus is used for DRAM on the main system board (motherboard) only, and is a multiplexed version of the system address bus. It uses 9 lines, MA0–MA8 which can be demultiplexed to provide the row and column address and strobe signals for DRAM. The X address bus is obtained from the system bus, and is used for addressing ROM BIOS and ports located on the main system board. The X address lines are obtained from the 82231, and these lines are used also to generate addresses for DMA and memory refresh. The L address bus is intended for expansion boards, and consists of buffered but unlatched signals for A17–A23.

The local data bus consists of the D0–D15 lines from the CPU, and after buffering becomes the system data bus which interfaces with all other data buses under the control of the 82230 and 82231 chips. The memory data bus is 16-bits wide and is used for both ROM and DRAM. It is connected to the system data bus by way of buffers. The 8-bit X-data bus is intended for port use, such as DMA control, interrupt control, keyboard control and real-time clock.

The 82230/82231 will allow up to 16M of memory to be controlled, the maximum that is permitted by the 24 address lines of the 80286 or 80386SX CPUs. The chips automatically insert one wait state for memory access and four wait states for port actions, but these default numbers of wait state can be increased or reduced as required. Note that ROM access is usually significantly slower than RAM, so that some computers on switch-on copy ROM data into RAM ('shadowed ROM') in order to gain faster access to ROM routines.

Another chip, the 82335, is intended specifically to allow the 80386SX processor to be used in an AT bus system, and is intended to be used along with the 82230/82231 set. The internal block diagram is illustrated in Figure 7.29; the packaging is as a 132-pin plastic and quad flatpack. The DRAM controller of this chip is designed to be used with the 80386SX running at 16 MHz, with paged interleaved memory implemented using 100 ns DRAM with no wait states. Some functions overlap those of the 82230/82231 set, but are more specifically aimed at the 80386SX design. In particular, the 80325 converts 80386SX bus cycles into cycles that are compatible with 80286 systems.

The built-in DRAM controller allows the address-

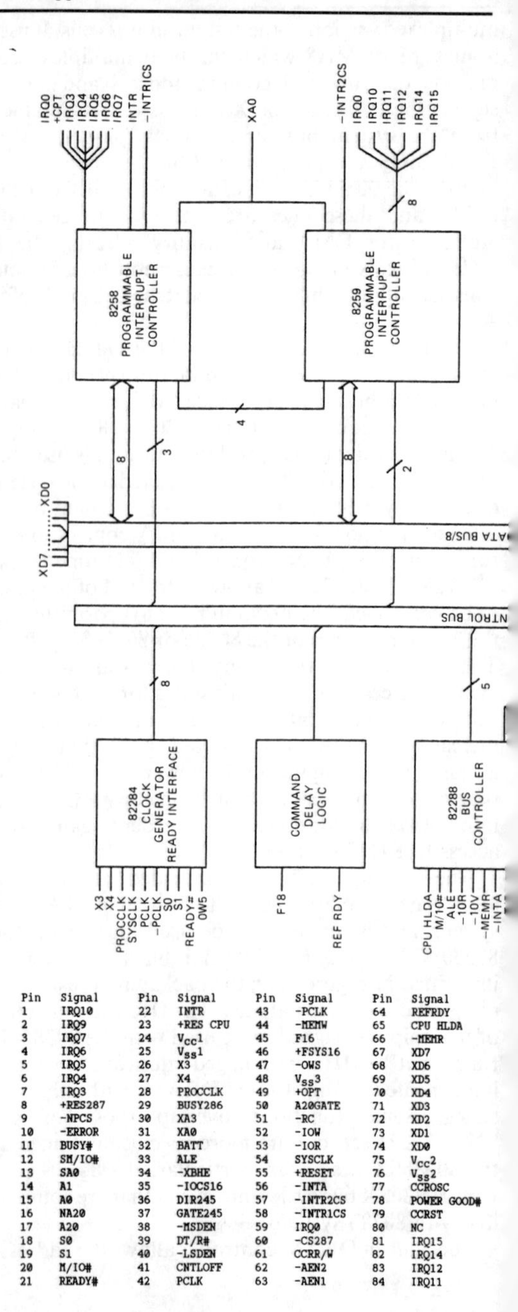

Pin	Signal	Pin	Signal	Pin	Signal	Pin	Signal
1	IRQ10	22	INTR	43	-PCLK	64	REFRDY
2	IRQ9	23	+RES CPU	44	-MEMW	65	CPU HLDA
3	IRQ7	24	$V_{cc}1$	45	F16	66	-MEMR
4	IRQ6	25	$V_{ss}1$	46	+FSYS16	67	XD7
5	IRQ5	26	X3	47	-OWS	68	XD6
6	IRQ4	27	X4	48	$V_{ss}3$	69	XD5
7	IRQ3	28	PROCCLK	49	+OPT	70	XD4
8	+RES287	29	BUSY286	50	A20GATE	71	XD3
9	-NPCS	30	XA3	51	-RC	72	XD2
10	-ERROR	31	XA0	52	-IOW	73	XD1
11	BUSY#	32	BATT	53	-IOR	74	XD0
12	SM/IO#	33	ALE	54	SYSCLK	75	$V_{cc}2$
13	SA0	34	-XBHE	55	+RESET	76	$V_{ss}2$
14	A1	35	-IOCS16	56	-INTA	77	CCROSC
15	A0	36	DIR245	57	-INTR2CS	78	POWER GOOD#
16	NA20	37	GATE245	58	-INTR1CS	79	CCRST
17	A20	38	-MSDEN	59	IRQ0	80	NC
18	S0	39	DT/R#	60	-CS287	81	IRQ15
19	S1	40	-LSDEN	61	CCRR/W	82	IRQ14
20	M/IO#	41	CNTLOFF	62	-AEN2	83	IRQ12
21	READY#	42	PCLK	63	-AEN1	84	IRQ11

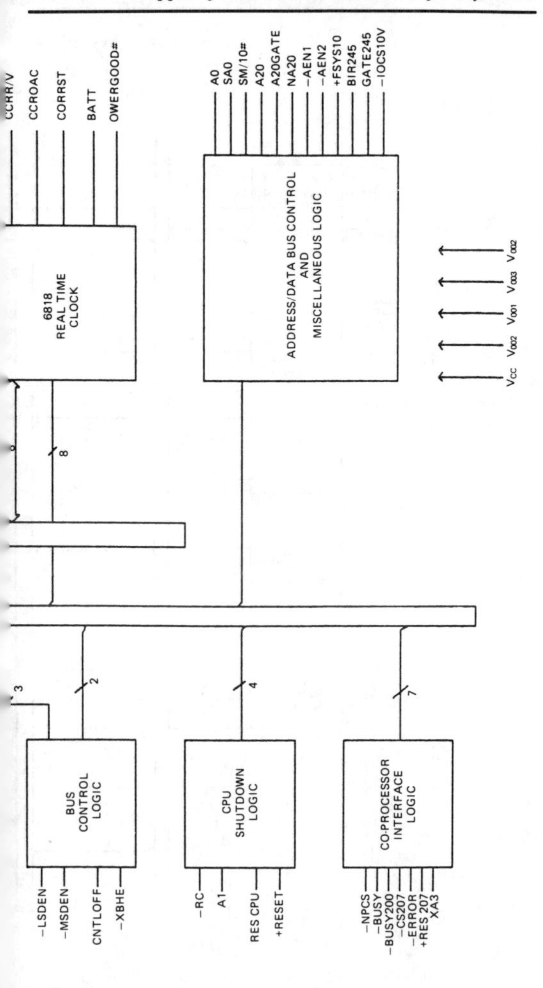

Figure 7.27 *The block diagram (a) and pin assignment (b) for the 82230 VLSI interfacing chip.*

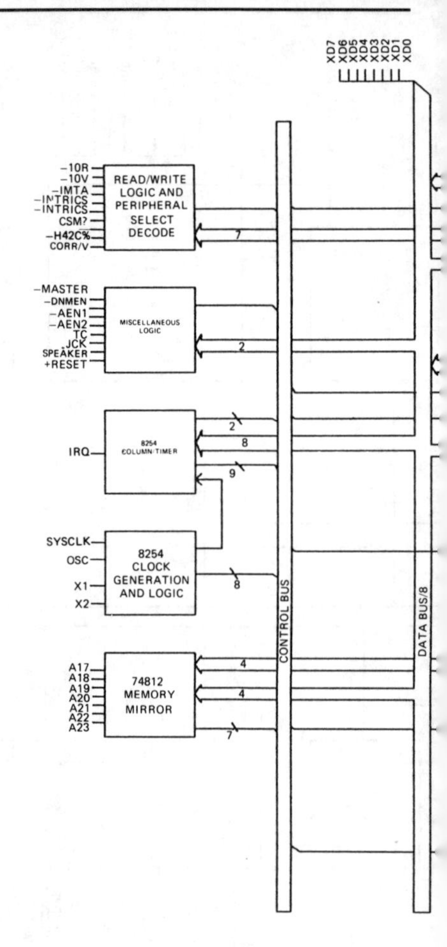

Pin	Signal	Pin	Signal	Pin	Signal	Pin	Signal
1	XA8	22	REFRDY	43	-IOCHCK	64	CPU HRQ
2	XA7	23	-AEN1	44	-XMEMR	65	NMI
3	XA6	24	-AEN2	45	-XMEMW	66	SPEAKER
4	XA5	25	-CCRR/W	46	DRQ0	67	OSC
5	XA4	26	-CS287	47	DRQ1	68	X1
6	XA3	27	IRQ0	48	DRQ2	69	X2
7	XA2	28	-INTR1CS	49	DRQ3	70	A23
8	XA1	29	-INTR2CS	50	DRQ5	71	A22
9	XA0	30	-INTA	51	DRQ6	72	A21
10	$V_{ss}1$	31	+RESET	52	DRQ7	73	A20
11	$V_{cc}1$	32	SYSCLK	53	$V_{cc}2$	74	A19
12	XD0	33	-IOR	54	$V_{ss}2$	75	A18
13	XD1	34	-IOW	55	-DACK7	76	A17
14	XD2	35	P2	56	-DACK6	77	XA16
15	XD3	36	+ACK	57	-DACK5	78	XA15
16	XD4	37	-DMAAEN	58	-DACK3	79	XA14
17	XD5	38	-RDXDB	59	-DACK2	80	XA13
18	XD6	39	-DPCK	60	-DACK1	81	XA12
19	XD7	40	-MASTER	61	-DACK0	82	XA11
20	NC	41	-REFRESH	62	T/C	83	XA10
21	HLDA	42	IOCHRDY	63	-8042CS	84	XA9

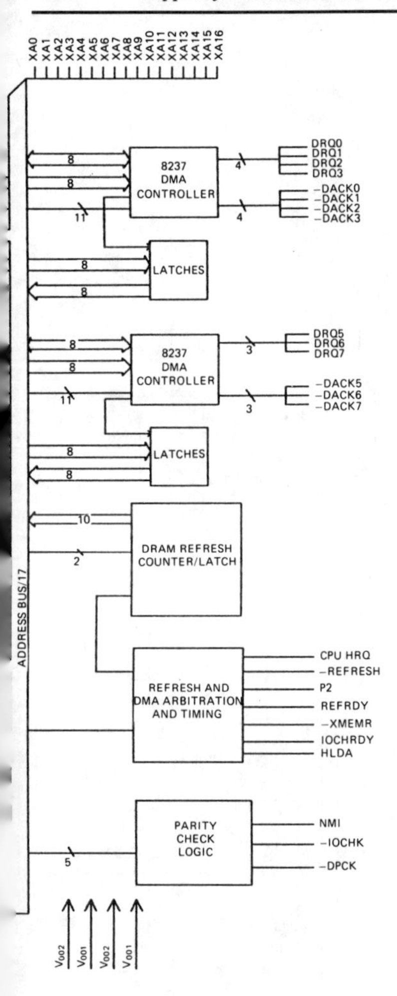

Figure 7.28 *The block diagram (a) and pin assignment (b) for the 82231 VLSI interfacing chip.*

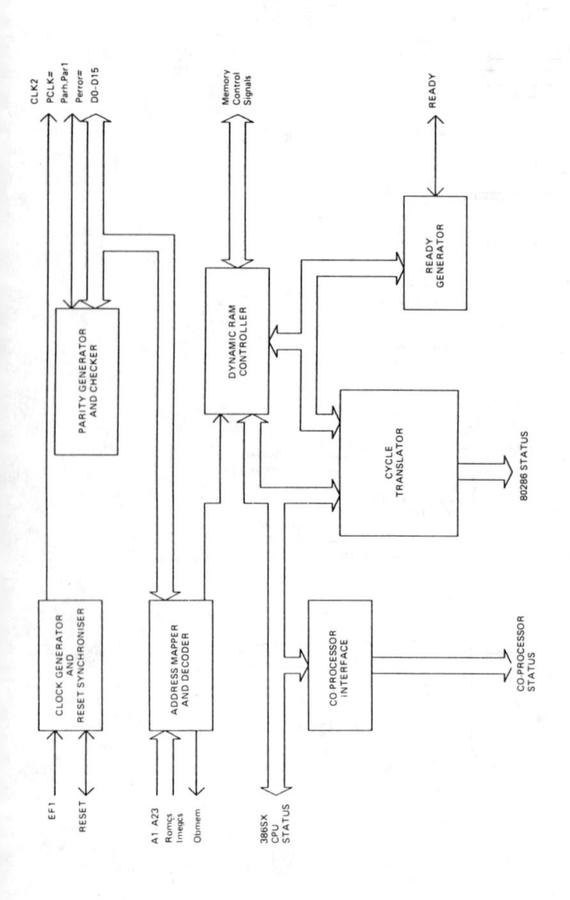

Figure 7.29 *The block diagram for the 82335 interfacing chip which creates the 'AT bus' used in most computers of the PC/AT type, and used in this example for a computer based on the 80386SX chip.*

ing and refreshing of up to four banks of 16-bits of DRAM, using either 256K or 1M memory chips. The 16-bit memory is, in fact, 18 bit since in the AT design 1 bit in each 8 is used for parity checking. The memory configuration is determined from a memory autoscan routine contained in the ROM, and once this has been determined at switch-on, the address lines A1–A23 are decoded to provide row and column address and strobe signals for DRAM use.

The MCA chip set

When IBM announced its PS/2 range of machines to replace the older PC types, one feature was the use of a new bus system to replace the AT bus. This system was designated micro-channel architecture (MCA), and is incompatible with the expansion boards that have been used so widely in the PC/XT and PC/AT designs. The 82310 and 82311 chip sets both provide for MCA use, but the 82311 contains more chip functions (Figure 7.30). The micro-channel bus consists of the address, data, arbitration and control bus lines of the system, using four-phase clocking of the 80386 or 80386SX processor. Chip sets are available for the clock speeds of 16 MHz, 20 MHz and 25 MHz, and 33 MHz versions are also obtainable.

Chip No.	Description	Included in - 82310	82311
82303	Local I/O support		√
82304	Local I/O support		√
82306	Local channel support	√	
82307	DMA/MCA arbitration control	√	√
82308	Microchannel bus controller	√	√
82309	Address bus controller	√	√
82706	VGA graphics controller	√	√
82077	Floppy disk controller		√

NOTE: Chip sets are available for 80386DX systems up to 25 MHz and for 16 MHz 80386SX systems.

Figure 7.30 *The chips contained in the two standard bus chip sets.*

The EISA chip set

The chip set for implementing the EISA form of bus, which allows a 32-bit bus to take advantage of a mixture of 32-bit, 16-bit and even 8-bit expansion cards, consists at the time of writing of four chips, with others yet to be delivered. Only provisional information was available at the time of writing. The four chips are the 82352 EISA bus buffer, the 82355 bus master interface, the 82357 Integrated System Peripheral and the 82358 EISA Bus Controller. These chips are 32-bit only and cannot be used by the 80286 or 80386SX CPUs, only by the 80386 and 80486.

Other system chips

This section has dealt with the support chips that are most closely associated with the CPU, and the other chips that are likely to be used in a computer system are simply listed below with brief comments.

Memory controllers:
8203 64K DRAM controller
8206 Error detection and correction unit
8207 Dual-port DRAM controller
82C08 RAM controller
Support peripherals:
8231A Arithmetic processing unit
8253 Programmable interval timer
8254 Programmable interval timer
82C54 CHMOS Programmable interval timer
8255A Programmable peripheral interface
82C55A CHMOS Programmable peripheral interface
8256AH Multifunction CPU support controller
8279 Programmable keyboard/display interface
82389 Message passing co-processor for MULTIBUS
Floppy and hard disk controllers:
8272A Single/double density FD controller
82077 Single chip FD controller
82064 CHMOS hard disk controller
Graphics co-processors:
82706 Intel video graphics array
82716 Video storage and display device
82786 CHMOS graphics co-processor

8 Addressing modes

I 16-bit processors (8088, 8086, 80188, 80186, 80286)

These processors allow for a total of eight modes of address for word or byte data. When data is contained in a data segment, each address will consist of two 16-bit numbers, the segment base number and the offset number. The segment base number can be implied (usually as DS) in the instruction, or it can be explicitly supplied by using a segment **override** such as DS, ES, CS, SS in the instruction.

The offset is the sum of up to three numbers:

- a displacement which is an immediate value of 8 or 16 bits contained in the instruction;
- a base, which will be the content of the BX or BP registers;
- an index, which will be the content of the SI or DI index registers;

and of these the base and the index will be used only in specialized addressing modes. When more than one number contributes to an offset, the sum is a 16-bit addition, with any carry **ignored**. Eight-bit numbers are treated as 16-bit numbers (so that the sign bit will be shifted if necessary). The resulting displacement number is also known as the **effective address**.

When a complete address number is formed, the segment number is shifted by 4 bits left and the offset (the effective address) is added to give the complete 20-bit address in memory. Software such as compilers must deal with any changes of segment that are required when code or data extends from one segment to another.

The implicit use of segments is as follows:

Code segment is used for instructions and immediate data.

Stack segment is used for PUSH and POP and for any memory reference which uses BP as the base register.

Extra segment is used for all string instructions which use the DI register as an index. Data segment is used for all other data references.

In most applications, multiple code segments will be used, and possibly multiple data segments. In the examples of addressing methods that follow, the segment is always implied.

Register and immediate addressing

Register operand addressing makes use of a register, either 8 or 16-bit, to contain the operand.

Examples: MOV AX, CX will copy the word from the CX register into the AX register. MOV BL, DH will copy the byte from the DH register into the BL register.

Immediate addressing uses the byte or word following the instruction and contained in the code segment immediately following the instruction.

Examples: MOV AX, 3FFFH copies the word 3FFFH into register AX. MOV AL, 5BH copies the byte 5BH into the AL register.

Memory reference addressing

Direct addressing makes use of a displacement word or byte to provide the offset. The displacement is contained in the instruction. The current data segment will be used unless overridden.

Examples: MOV AX, datwrd will load AX with the word beginning at location **datwrd**. This means that AL will contain the byte at location **datwrd** and AH will contain the byte located at **datwrd** + 1. MOV AX, [3FFFH] will load AX with the word in locations 3FFFH, 4000H – note that the square brackets are needed to indicate that 3FFFH is not an immediate number. Different assemblers may treat this distinction in different ways.

The use of an 8-bit register for direct addressing automatically implies that only 1 byte will be loaded, so that MOV AL, **datbyt** will copy the byte at location **datbyt** into register AL and nothing will be loaded into AH. Register indirect addressing uses also a single displacement byte or word which is contained in one of the registers SI, DI, BX or BP.

This allows the instruction to be much more compact, and also permits the use of loop actions in which the number in the register is incremented or decremented on each pass.

Examples: MOV AX, [BX] will load the AX register with the word copied from the memory address which is contained in the BX register. This implies that AL is loaded from the address in BX and AH is loaded from the following byte address. As before, if a byte register is used as the destination, only 1 byte will be loaded.

Base addressing makes use of an offset which is contained by adding a displacement (byte or word) to the contents of BX or BP (ignoring any carry). The operand must specify both register and displacement.

Example: MOV AX, [BP+2FFH] will add the contents of the BP register to the number 2FFH and use this resulting number as the offset within the current data segment. As before, specifying an 8-bit register as destination will result in 1 byte being copied.

Index addressing makes use of the index registers SI or DI. The offset is the result of adding the number in the index register to the number which is supplied as a displacement. The number is provided as a label name which is usually the location of the first byte of an array.

Example: MOV AX, list[DI] will add the memory address list to the content of DI and use the resulting number as an offset. For reading words in an array, DI would start by holding 0, and would be incremented to 2, 4, 6, etc. for subsequent loads in a loop.

Based indexed addressing uses an offset number which is obtained by summing the contents of a base register (BX or BP) and the contents of an index register (SI or DI), disregarding any carry from the addition.

Example: MOV AX, [DI+BX] will sum the contents of the DI and BX registers and use the resulting sum as the offset within the data segment to locate a word in the memory to load into AX.

This instruction can also be written as MOV AX, [DI] [BX].

Based indexed addressing with displacement uses an instruction which specifies a displacement as well as the index register and the base register.

Example: MOV AX, [DI + BP + 0FH] will sum the contents of DI and BP with the immediate number 0FH, disregarding any carry, and use the result as an offset in the data segment to locate a memory address from which a word is loaded into AX.

I/O port addressing

The I/O address space consists of 64K of 8-bit ports or 32K of 16-bit ports. An 8-bit address can be specified as an immediate number, using an instruction of the forms:

IN AX, 00F0H (load word from port 00F0H)

or

IN AL, 00F0H (load byte from port 00F0H)

and when the 8-bit port address is used, the bits A8–A15 will all be zero. For 16-bit port addresses, the DX register is used to hold the port address and the instructions are of the form:

IN AX, DX

or

IN AL, DX

according to whether a word or a byte is to be loaded. The corresponding output instruction is OUT, taking the same operands.

The port addresses 00F8H to 00FFH are reserved and must not be used by software.

II 32-bit processors (80386DX and 80386SX)

So that compatibility can be preserved, the addressing methods of the 32-bit processors follow the same scheme as for the 16-bit processors, with some enhancements which are listed here.

1 The registers allow for operands that are byte, word or double-word (Dword) size. The usual AL, AX type of distinction is used for the byte and word actions, with the prefix **E** used to indicate an extended 32-bit register, so that any addressing instruction that uses register names such as EAX, EBX, ECX and so on refers to 32-bit transfers.

2 In addition to the address components of displacement, index and base, the 32-bit processors allow a **scale factor** which can be 1, 2, 4 or 8. Adding a scale number allows an indexed addressing method to work in units of byte, word, D-word or quadword (8 bytes). In a scaled addressing method, the content of the index register is multiplied by the scale factor before adding a displacement and/or base register contents.

3 In real mode, including virtual 8086 mode, all operand lengths and effective address numbers are assumed to be 16-bits long. In protected mode, the default length is 32-bits. These defaults can be overridden for an individual instruction by using an operand size prefix or an address length prefix. Globally, the difference is achieved by using the D-bit in the CS segment descriptor.

4 I/O space still consists of 64K bytes, but this can be allocated to 64K of 8-bit ports, 32K of 16-bit ports or 16K of 32-bit ports. The usual 00F8H to 00FFH port addresses are reserved.

9 Instruction sets

I Original 16-bit chips (8088, 8086, 80188, 80186)

The 8088 and 8086 chips use an instruction set which can be classed into 18 groups of instructions, most of which can make use of any of the available addressing methods which are relevant to that instruction. Since all other chips of the Intel family are upwards compatible, this forms a base set that can be used for software that can then be executed by any chip in the family. The instruction groupings, using the Intel nomenclature, are shown below with the assembler instruction words.

General-purpose

MOV	Move word or byte from source to destination
NOP	No operation
POP	Pop word from stack
POPA	Pop all registers from stack
PUSH	Push word on to stack
PUSHA	Push all registers on to stack
XCHG	Exchange byte or word between registers
XLAT	Use translate table to load byte

Input/output

IN	Input byte or word from port
OUT	Output byte or word from port

Address object

LDS	Load pointer to DS and specified register
LEA	Load effective address of label name
LES	Load pointer to ES and specified register

Flag transfer

LAHF	Load AH register from flag register

POPF	Pop flags from stack
PUSHF	Push flags on to stack
SAHF	Store AH register contents to flag register

Addition

AAA	ASCII adjust for addition
ADC	Add byte or word with carry in
ADD	Add byte or word, no carry in
DAA	Decimal adjust for addition
INC	Increment byte or word by 1

Subtraction

AAS	ASCII adjust for subtraction
CMP	Compare byte or word
DAS	Decimal adjust for subtraction
DEC	Decrement byte or word by 1
NEG	Negate (invert) byte or word
SBB	Subtract byte or word with borrow
SUB	Subtract byte or word, no borrow

Multiplication

AAM	ASCII adjust for multiply
IMUL	Integer multiply byte or word
MUL	Multiply byte or word unsigned

Division

AAD	ASCII adjust for division
CBW	Convert byte to word
CWD	Convert word to doubleword
DIV	Divide byte or word unsigned
IDIV	Integer divide byte or word

Strings

CMPS	Compare byte or word string
INS	Input byte or word string
LODS	Load byte or word string
MOVS	Move byte or word string
OUTS	Output byte or word string
REP	Repeat string action
REPE	Repeat string action while equal
REPNE	Repeat string action while not equal
REPNZ	Repeat string action while not zero
REPZ	Repeat string action while zero
SCAS	Scan byte or word string
STOS	Store byte or word string

Logical actions

AND	AND byte or word

NOT	Negate (invert) byte or word
OR	OR byte or word
TEST	Test byte or word
XOR	XOR byte or word

Rotates

RCL	Rotate left through carry, byte or word
RCR	Rotate right through carry, byte or word
ROL	Rotate left, byte or word
ROR	Rotate right, byte or word

Shifts

SAL	Shift arithmetic left, byte or word
SAR	Shift arithmetic right, byte or word
SHL	Shift logical right, byte or word (identical to SAL)
SHR	Shift logical right, byte or word

Flag actions

CLC	Clear carry flag
CLD	Clear direction flag
CLI	Clear interrupt enable flag
CMC	Complement carry flag
STC	Set carry flag
STD	Set direction flag
STI	Set interrupt enable flag

Synchronization

ESC	Escape to extension processor
HLT	Halt until interrupt or reset
LOCK	Lock bus during next operation
WAIT	Wait until TEST pin active (LOW)

Interrupts

INT	Interrupt
INTO	Interrupt if overflow
IRET	Return from interrupt

Loops

JCXZ	Jump if CX register contains zero
LOOP	Loop unconditionally
LOOPE	Loop if equal
LOOPNE	Loop if not equal
LOOPNZ	Loop if not zero
LOOPZ	Loop if zero

Unconditional jumps

CALL	Call procedure/subroutine
JMP	Jump to new address

RET	Return from procedure/subroutine

Conditional jumps

JA/JNBE	Jump if above/not below or equal
JAE/JNB	Jump if above or equal/not below
JB/JNAE	Jump if below/not above or equal
JBE/JNA	Jump if below or equal/not above
JC	Jump if carry set
JE/JZ	Jump if equal/zero
JG/JNLE	Jump if greater/not less nor equal
JGE/JNL	Jump if greater or equal/not less
JL/JNGE	Jump if less/not greater nor equal
JLE/JNG	Jump if less or equal/not greater
JNC	Jump if carry clear
JNE/JNZ	Jump if not equal/not zero
JNO	Jump if no overflow
JNP/JPO	Jump if not parity/parity odd
JNS	Jump if sign bit clear
JO	Jump if overflow bit set
JP/JPE	Jump if parity bite set/parity even
JS	Jump if sign bit set

Note: The 80188 and 80186 contain a further group of **high-level** instructions which are:

BOUND	Detect values outside prescribed range
ENTER	Format stack to enter procedure
LEAVE	Restore stack for leaving procedure

II The 80286 Chip

The 80286 chip permits the use of all groups of instructions specified for the 8088/8086 chips, and adds the high-level instructions of the 80188/80186, plus a set of execution environmental controls. The differences and additions are as follows:

1 The WAIT instruction in the synchronization set waits for the BUSY pin to be inactive.

2 The execution environment control set consists of:

LMSW	Load machine status word
SMSW	Store machine status word

See Section 6 for a brief description of changing from real to protected mode.

III The 32-bit chips, 80386DX and 80386SX

The 32-bit chips can use all of the basic set of instructions of the 8088/8086, plus the enhancements of the 80188/80186 and 80286. To this is added several groups and individual instructions which are peculiar to the 32-bit chips.

The conversion commands of CBW and CWD which are shown in the division group of the 8088/8086 are considerably enhanced in the 32-bit chips, leading to a conversion group:

Conversion group

CBW	Convert byte to word (as for 8088/8086)
CDQ	Convert Dword to Qword (8 bytes)
CWD	Convert word to Dword (as for 8088/8086)
CWDE	Convert Word to Dword extended
MOVSX	Move byte or word, Dword, sign extended
MOVZX	Move byte or word, Dword, no extension

Single bit group

BSF	Bit scan forward
BSR	Bit scan reverse
BT	Bit test
BTC	Bit test and complement
BTR	Bit test and reset
BTS	Bit test and set

Protection model group

ARPL	Adjust requested privilege level
LAR	Load access rights
LGDT	Load global descriptor table
LIDT	Load interrupt descriptor table
LLDT	Load Local descriptor table
LMSW	Load machine status word
LSL	Load segment limit
LTR	Load task register
SGDT	Store global descriptor table
SIDT	Store interrupt descriptor table
SLDT	Store local descriptor table
SMSW	Store machine status word
STR	Store task register

VERR	Verify segment for reading
VERW	Verify segment for writing

Commands and timing

The range of 8086/8088 commands is very large if every possible combination of register and addressing method is considered, but the use of an assembler makes it completely unnecessary to know what codes will be generated from these differences. In this table, only a selection of basic commands is shown because this book is primarily concerned with the hardware of the Intel chips and in any case, some 90 per cent of program applications will use only a small subset like this.

Where no addressing is needed the words **opcode only** are added. Minimum and maximum times for commands are shown in terms of clock cycles. The timing of most commands depends on the type of addressing method that is used, and where the letters **EA** occur in a time, this means that a number of clock cycles has to be added to allow for the addressing method as follows:

Word operand at odd-numbered address	: 4 clocks
Immediate address	: 6 clocks
Base BX, BP, SI, DI	: 5 clocks
Base + displacement	: 9 clocks
BP + DI or BX + SI	: 7 clocks
BP + SI or BX + DI	: 8 clocks
(BP + DI or BX + SI) + displacement	: 11 clocks
(BP + SI or BX + DI) + displacement	: 12 clocks

Immediate addressing or register-to-register transfer requires the smallest number of clock cycles, memory to/from register the largest, **before** allowing for the addressing method. Only the range is indicated here, because you seldom have to allow for timing unless you are working with critical timing such as for serial communications. For details of times of instructions, see the Intel manuals for the 8086/8088 chip. Remember that operands must be compatible in size (byte or word), and that transfer from memory location to memory location (not involving registers) is not used – this is a DMA function.

The i486 can be programmed directly with the floating-point commands of the 80387 and other co-processor chips, so that the use of the i486 is in this respect equivalent to the use of the 80386DX with the 80387.

ADC	Add with carry. 3 clocks minimum, 17+EA clocks maximum – add integer with carry to integer in register or memory. Can use byte or word register/memory.
ADD	Add, no carry. 3 clock minimum, 17+EA maximum – add integers without carry from previous action, using register or memory.
AND	Logic AND. 3 clocks minimum, 17+EA maximum – AND byte or word in register or memory with byte or word in memory or register.
CALL	Call subroutine. 16 clocks minimum, 37+EA maximum – call subroutine whose address is given in operand, return to following instruction when completed.
CMP	Compare operands. 3 clocks minimum, 10 + EA clocks maximum – compare one operand in register or memory with another.
DEC	Decrement. 2 clocks minimum, 15+EA maximum – decrement stored number by unity.
IN	Input from port. 8 clocks minimum, 10 clocks maximum – read byte or word into AL or AX from numbered port.
INC	Increment. 2 clocks minimum, 15+EA maximum – increment stored number by unity.
INT	Interrupt (operand is number). 51–52 clocks.
IRET	Return from interrupt. Opcode only 24 clocks – return from interrupt to point of interruption.
J(type)	Jump conditional. 4–6 clocks if no jump, 16–18 if jump. – J is followed by letters to indicate condition, giving JA, JGE, JNE, JZ etc. Jump must be within limits of +127 and –128 places of operand.

JMP	Jump unconditional. 15 clocks within segment, up to 24+EA otherwise.
LOOP	Loop ending. 5 clocks minimum (end of loop), 19 maximum (loop back) – can be followed by condition like **E**, **NE**, **Z**, **NZ** or alone. CX is always tested as well as other conditions.
MOV	Move data. 2 clocks minimum, 10+EA maximum – move data to or from memory or register.
NOT	Complement number. 3 clocks minimum. 16+EA maximum – invert integer number (ones complement)
OR	Logical OR. 3 clocks minimum, 17+EA maximum – OR byte or word with byte or word in register or memory.
OUT	Port output. Operand is port number 8–10 clocks – output byte or word in AL or AX to port.
POP	Pop word from stack. 8 clocks minimum, 17+EA maximum.
POPF	Pop flags. Opcode only. 8 clocks – pop the stack word to the flags register.
PUSH	Push register. 10 clocks minimum, 16+EA maximum – push contents of named register to stack.
PUSHF	Push flags. Opcode only. 10 clocks – push the contents of the flags register to the stack.
RLC	Rotate left with carry. 2 clocks minimum, 20+EA+4*CL maximum – rotate register left, including carry, using one bit rotation, or the number of bits specified in CL.
RCR	Rotate right with carry. 2 clocks minimum, 20+EA+4*CL maximum – rotate register right, including carry, using one bit rotation or the number of bits specified in CL.
RET	Return. 8 clocks minimum, 18 maximum – return from subroutine (can also add to stack pointer).

ROL	Rotate left. 2 clocks minimum, 20+EA+4*CL maximum – rotate left, not using carry, by one bit or by number of bits specified in CL.
ROR	Rotate right. 2 clocks minimum, 20+EA+4*CL maximum – rotate right, not using carry, by one bit or by number of bits specified in CL.
SAL/SHL	Left shift. 2 clocks minimum, 20+EA+4*CL maximum – shift left, not including carry, by one bit or by number specified in CL. SAL and SHL are identical codes.
SAR	Arithmetic right shift. 2 clocks minimum, 20+EA+4*CL maximum – shift right, not including carry, by one bit or by number specified in CL.
SBB	Subtract with borrow. 3 clocks minimum, 17+EA maximum – subtract integers, using carry flag to mean a borrow.
SHR	Shift right logical. 2 clocks minimum, 20+EA+4*CL maximum – shift right by one bit or by number specified in CL.
SUB	Subtract. 3 clocks minimum, 17+EA maximum – subtract integers, byte or word.
XCHG	Exchange. 3 clocks minimum, 17+EA maximum – exchange contents of memory or register with compatible register.
XOR	Exclusive OR. 3 clocks minimum, 17+EA maximum.

10 Operating system calls and interrupts

DOS development

When microcomputers started to be used seriously for business purposes, many manufacturers devised their own operating systems. The aim of the operating system is to provide, at the very least, the essential routines that allow the computer to make use of the screen, the keyboard and, in particular, the disk drive(s), managing files and loading and running programs. Without program routines that attend to these tasks, the computer is unusable, so that the operating system is just as important as all the hardware of the computer. The use of disks is so central to any computer that will be used for serious purposes that the operating system (OS) is always referred to in its abbreviated form of DOS, disk operating system.

The use of different operating systems by different manufacturers, however, makes it virtually impossible to set standards that will allow one computer to run a program that works on another computer from a different manufacturer. When all computers were mainframes, with one computer to one company, programs were specially written for each machine and interchangeability of programs had no relevance. The prospect that each user of a computer might eventually be working with a separate machine came only with the development of microcomputers, and only when corporate users started to take microcomputers seriously.

One solution to the problem of interchangeability of programs came about in the very early days of the development of microcomputers, in the form of CP/M. This was a system to control, program and monitor (hence the initials) the operation of these early microcomputers, and it was devised by Gary Kildall in or around 1973, before microcomputers were a commercial proposition. At that time, the idea of a 'portable' operating system, one that could be used on more than one machine, was not new because Unix was already under development for mainframe machines, but the idea of a standard operating system for microcomputers was certainly new.

One particularly valuable feature of CP/M was that it came along with a set of utility programs that allowed the user to carry out tasks like file copying, file checking, program modification, disk verification and all of the other housekeeping actions that are now so familiar. Some of these programs had been devised in the very early days of microcomputer use, and were intended for the programmer rather than the user, so that they used only the minimum of screen messages and accepted single-letter commands that the user had to remember. They were, however, extremely efficient in the sense of providing a great deal of control in a package that took up very little space in the memory and on the disk.

The start of serious small-scale computing as we know it today was with the IBM PC machine, which was released in 1981. Before the machine could be released, however, it needed an operating system, one that would provide the facilities of CP/M on an extended scale for a complete new generation of 16-bit machines. Such an operating system should preferably be rather similar to CP/M in the way that it used commands (and some commands could be identical) so that users would have something familiar to hold on to rather than feel that everything had to be relearned from the start.

Digital Research, the company that had been set up to market and develop CP/M, offered a new version, CP/M-86, which could be used on the PC type of machine. At the same time, however, the rival software house Microsoft were using a 16-bit operating system which had been developed by a smaller firm, Seattle Computer Products in 1980, and which at one time was termed QDOS, with the QD allegedly meaning Quick and Dirty. This was renamed 86-DOS and with intensive effort was developed into a powerful operating system which was offered to IBM at a price that considerably undercut CP/M-86.

The system as used on the PC was known as PC-DOS, and a virtually identical version was available for any other manufacturers of machines using the

Intel chips, using the name MS-DOS. This has progressed through many versions (it has now reached version 5.0) as the demands for new facilities has expanded; the main changes in the later versions have been to allow for networking use and for use with the 80386 type of machine in which the memory limitations of earlier versions are overcome to a limited extent.

The main thrust of development, however, was to make a DOS which was considerably more extensive than CP/M, with a considerable redesign of fundamentals. Both CP/M and MS-DOS, for example, make extensive use of **registers** in the processor. A register is a form of memory store in which actions can be carried out on any byte or pair of bytes (a **word**) that is stored in the register. In CP/M, for example, one register can be loaded with a byte that is to be written to the disk. Another register is then loaded with a command byte, and when a command is executed to call CP/M, the byte is written. When this is done, the contents of these registers are usually changed, so that the action cannot be repeated without reloading all of the registers. MS-DOS is written so as to ensure that these contents are not changed unless they have to be, so that using MS-DOS for writing software is considerably more straightforward. Another point is that commands are made more rational. In CP/M, the copying command is PIP, and to copy a file **TESTFIL** in drive A onto Drive B needs the command in the form:

PIP B:=A:TESTIT

– with the destination written first. The corresponding MS-DOS command uses the word COPY and in the form:

COPY A:TESTIT B:

which seems more logical in form and arrangement, particularly to the owner who is more interested in using the commands than with their design.

The first IBM PC machine used only 16K of RAM, and in its standard form used an external audio cassette recorder for storage rather than disks. Disk drives could be added, however, with a memory extension to 64K, which seemed adequate at the time, since all of the standard business programs (such as WordStar, SuperCalc, dBase-2) ran under

CP/M in this amount of memory space on the CP/M machines and could be converted to run in the same amount of space on the new machine. The 8088 chip and its stablemate the 8086 allowed up to 1 Mb of memory to be used, however, so that pressure soon built up for the use of more memory, particularly when the XT machine, which could use a hard disk, was launched.

As the demand for the PC grew, and the disk systems for other machines were standardized to the familiar 5.25 in. 40-track double-sided double-density format, MS-DOS was being steadily developed to cope with new needs. One restriction that has remained on MS-DOS is one of memory management. MS-DOS was originally written to cater for a maximum of 10 segments of RAM, corresponding to 640K, and this has to date never changed, though version 4.0 allows for the use of both extended and expanded memory (see later for explanation) and V5.0 provides for larger memory sizes to be used. The limit is fixed by the use of a 1 Mb address limit on the original 8088/8086 chips, which both use a 20-line address bus, so that the address range covers 2^{20} addresses. This is 1 048 576, which is one megabyte (1 Mb), allowing 640K for RAM and the remaining 384K for ROM and for RAM that is used for the screen display. In addition, MS-DOS was conceived from the start as a single-user single-tasking DOS. This means that it is intended for a machine used by one person and running one program. A multi-user system permits other users with keyboards connected to the machine to run programs, and a multitasking system implies that more than one program can be run at one time, with the user being able to switch from program to program.

The development of higher-resolution screen displays from the original text only, through monochrome graphics to colour graphics of the GCA, VGA and EGA variety, has made increasing inroads into the 348K part of the memory allowance, and a further piece has been taken by the hard-disk ROM (whether the hard disk is built-in or added as a card). The later 80286 chip was constructed in a way that allowed it to handle up to 16 Mb of memory – but only when running a different operating system such as Xenix or Unix, and it retained the 1 Mb limit for

MS-DOS. Only the 80386 chip, of those announced and being delivered to date, overcomes the problems of 64K segments and the 1 Mb memory limit – but MS-DOS has not adapted fully to this freedom. Later versions of MS-DOS were adapted for networking, allowing computers to be connected together in order to share a hard disk drive and a printer, but provision for multitasking use has been confined to add-on programs (like Windows).

It is true that the later version, MS-DOS 4.0, has provision for using add-on memory, the type called **expanded** memory, as used on XT machines, which is added in the form of a card, and switched into circuit as required. This is done by using part of the unused memory space between 768K and 960K as a 64K 'window' which can be mapped on to additional RAM, though this implies that only 64K of additional RAM is available at any one time, and a suitable device driver must be present and loaded by way of the CONFIG.SYS file.

This system should not be confused with the **extended** memory which can be added to 80286 and 80386 machines. These machines use a 24-line address bus so that 16 Mb can be addressed, and memory that is installed beyond the original 1 Mb limit on such machine is extended memory, which can be used only by programs that have been written to do so.

Neither form of added memory can be used, at present, in any simple way. MS-DOS therefore does not provide for multitasking when this 640K requirement is exceeded. The earlier versions of the MS Windows system allow for more than one program to be present, and for switching easily between programs, but all within the 640K limit, using a hard disk to store a program and data while the use of the program is suspended. Windows allows for true multitasking only on the 80386 type of machine (using Windows-3.0), and the 640K barrier is broken by the later operating system, OS-2 rather than by MS-DOS. However, OS-2 requires very much more memory for its own purposes, and is intended for machines with many megabytes of RAM memory, and it is possible that later versions of MS-DOS, beyond 4.1, may offer multitasking without being so demanding on memory. Another 'front-end' program, DesqView, can be obtained in both 286 and 386 versions and allows a range of multitasking activities.

DOS versions

The original versions of MS-DOS were 1.0 and 1.1. These were suited to the older version of the PC, and V.1.0 could be used only with single-sided disk drives. V.1.1, which allowed the use of the familiar double-sided disk, was released in 1982. These early versions were rare in Europe, where the PC type of machine was not outstandingly popular at first.

Versions 2.0 and 2.1 were written to expand the use of MS-DOS very considerably when the PC was almost universally supplied with a disk system and more memory, often with 360K or more. At this time, clones of the PC began to appear, some of which provided as standard many of the facilities that had been available on the original machine only by way of plug-in cards. The most important of the additions that were made to MS-DOS at that time were to cope with TSR programs, country-variables, hard disk use, and file handling.

A terminate-and-stay-resident program (TSR) is one that can be loaded into a segment and which will then stay in that segment, being activated and deactivated as needed within a computing session without the need to be reloaded. The use of such programs involves the allocation of memory and the protection of that memory in order to make sure that no other program is loaded over the TSR program.

There also must be provision for 'hot-key' use, so that a key combination, such as alt-shift can start or stop the TSR program. What is not usually provided for is the removal of a TSR program cleanly from the memory so that the memory can be released for other use. This often means that loading in a large program, such as a desk-top publishing package or a large program like Lotus Agenda or Symphony can lead to *out of memory* messages, and the program can be run successfully only if the machine is rebooted first in a way that does not load the memory-resident utilities – this usually means rewriting the AUTOEXEC.BAT file. There are utilities for freeing memory, such as HEADROOM, but few of the utilities that are used as TSR programs have any provision for releasing the memory – one honourable exception is Logitech's CATCH. The Public Domain Software Library lists the memory-releasing utilities RELEASE and FIX27 – see the address in Appendix D.

Country variables refer to the keyboard and screen characters that will appear in versions of MS-DOS that run on machines which are used with a character set other than the standard US one. This is provided for in DOS versions from 2.0 on, but versions 2.0 and 2.1 provide only for getting this information, not for easily changing it. The ability to view and modify the country variables was completely implemented in version 3.0.

The provision for hard disk drives was a particularly important step forward. This was made necessary because of the fast rate of development of miniature hard disks from the 8 in. size to 5.25 in. and then to 3.5 in., making it possible to have a hard disk in the space of a floppy drive, or even on a card that would fit into one of the slots inside the computer. The familiar commands that were added at this time include MKDIR (or MD) to make a disk directory, CHDIR (or CD) to switch to a named directory, and RMDIR (or RD) to remove a directory. The drive memory limit was set at 32 Mb, after which the disk drive had to be **partitioned** as if it were more than one drive.

At the same time, MS-DOS 2.0 added a large number of routines for handling files, most of which are not accessible directly through command names but which can be used by programmers working in assembly language or in some higher level languages, notably C. The EXIT command was also added, which makes it possible to jump out of a program (the **parent** program) to use DOS (as a **child program**), and then (by typing EXIT and then pressing the RETURN key) to return to the program. Only a few programs are written in order to take advantage of this.

Another major addition was a supplementary way of controlling files. The earlier versions of MS-DOS had used the same method of keeping track of files as CP/M, a method called file control blocks (FCB). With MS-DOS 2.0, this older method was supplemented by a set of new routines which could use a much simpler method, a file handle. Without going into detail, using a file control block meant that a set of bytes in memory had to be used to hold, in correct order, all the details of a file, and this form of memory block had to be set up for each file.

The file handle method allowed a simple string of ASCII characters, consisting of drive letter and full filename, to be used as a file control by assigning a 16-bit code number, with all the rest of the work performed by the operating system. The file-handle system also allowed peripherals like the printer and the serial port to be used as if they were files, with names such as PRN and AUX1. Programs could still be written using the older method, because it was still available on DOS 2.0 and beyond, but the file-handle method is so much easier to use that most programmers have taken advantage of it.

There were also some useful minor additions in MS-DOS 2.0, such as the ability to control the use of the break key, and the ability to rename a file and to label a file with date and time. Also added was the information on the space remaining on a disk (following the disk directory) and the VER command to get the MS-DOS version number.

The large changes between versions 1.1 and 2.0 or 2.1 meant that programs that were written to take full advantage of version 2.0 or 2.1 could not be used with 1.0 or 1.1. This incompatibility was less serious that it seemed, because the major changes that had taken place in the requirements for DOS, such as hard disk use, made it pointless to write programs that would use the older version, and it was still possible to run programs that had run under the older versions. The changes that were made from version 2.1 to version 3.0 (followed in quick succession by 3.1, 3.2, 3.3 and 3.4) were not so dramatic from the point of view of the user of a single machine, but they added significant extensions.

The main extensions concerned the use of programs by computers that were part of a network. Networking requires changes to the operating system in order to make certain that one computer does not, without permission, alter the files that belong to another, and quite extensive additions to MS-DOS had to be made in order to satisfy the requirements. For users of solo machines, however, these had no impact, but the addition of pipes and filters did.

The principles of pipes and filters had existed for a considerable time in the Unix operating system for mainframe computers, but they had not previously been implemented in an operating system for a microcomputer. A pipe implies that a command symbol will allow data to be channelled from one program to another, or from a program to a different outlet (to printer, for example, rather than to screen). A filter means that a program which can modify data can be placed in the way of data that is being transferred. For example, data that is to be shown on the screen can have a filter inserted that will ensure that it appears in alphabetical order, or it can be arranged to appear 24 lines at a time so that the user has time to look at one screen before seeing the next 24 lines. Redirection of this type was the main step forward in version 3.0 as far as the single user was concerned.

As a more minor change, the country variables could now be set by the use of KEYB files, such as KEYBUK which sets the keyboard for the UK character set, with the pound sign, and with the single and double quote marks in positions that are transposed in comparison to the US keyboard.

Version 3.2 also introduced support for 3.5in. disk drives, allowing a change to this size of disk. This was rather belated as far as the PC was concerned, because machines such as the Apple Macintosh, Amiga and Atari ST had shown the advantages of the smaller and better-protected disks.

In general, most users of MS-DOS should by this time be using version 3.0 at the earliest, and many will have obtained version 3.3 or 3.4 with their computers in recent years, though very few at the time of writing have V.4.0 or later. The differences between the various subversions since 3.0 are less important than the step from 2.2 to 3.0, and it is not advisable to run any modern program on version 2.1 or earlier.

Version 4.0

Version 4.0 became available in 1988, and seems to be intended as a way of preparing users for the use of OS/2 (which has been long delayed). The 32 Mb limit on the hard disk drive was removed, so that partitioning is no longer needed, and at last there was some use of memory above the 640K limit. The main difference for the user, however, is that the screen presentation has changed to the windows/icon/mouse style, with the inevitable sacrifice of memory that this entails.

V.4.0 is contained as standard on two 3.5 in. disks since it will be supplied mainly with machines that use 3.5 in. disks, though if any manufacturer intended to use MS-DOS 4.0 on a machine that used 5.25 in. disks the program would be available on these disks. One disk is labelled as an Install disk, and is used to boot in the operating system in the usual way. Normally, DOS 4.0 will be supplied only with a new machine whose hard disk will be unconfigured, but if you change to V.4.0 from an older version such as 3.3, then you can expect to run into trouble if your hard disk is already partitioned. Some users have, at the time of writing, reported incompatibility problems with DOS 4.0 when running older programs.

Another novelty is the automatic creation of CONFIG.SYS and AUTOEXEC.BAT files, along with a new file, DOSSHELL.BAT which controls other aspect of use, such as mouse support and the command display. This now provides the option of using command prompt, which means that programs can be run in the same way as on older versions, by typing the program name (and drive/directory-path if needed) or by using the file system which allows selection by mouse.

The amount of extra memory needed for these systems, however, means that some programs, notably large graphics programs like Lotus Freelance Plus (see Section 7), will not have enough memory to run, even if expanded memory is present to be used by DOS. The original version of Microsoft Windows 386 will not work with DOS 4.0, though a later version 2.10 will do so. On the whole, since Microsoft are reluctant to release any details of V.4.0 other than to computer manufacturers, it is better to use DOS 4.0 on computers that come with the system than to attempt to upgrade on your present machine. It is also important to check that all of your software is compatible with DOS 4.0 (or DOS 4.01), because some problems in this respect have been reported with the earlier issues of DOS 4.0.

V5.0 has been announced for late 1990 and is reported as fixing problems in V4.0 along with better management and the use of considerably less memory than older versions.

COM and EXE files

MS-DOS can run two types of machine-code program which use the extension filenames of COM and EXE respectively. A COM file is constructed almost identically to a CP/M file (which would also carry the COM extension), and the entire program, along with all of its data and other use of memory, is contained in a single 64K segment. MS-DOS will decide which segment is selected (always the lowest-numbered part of the memory), and because all of the program is contained within a 64K address space the address numbers can use single-word (16-bit) registers. This make it possible to design and use short and memory-efficient programs, and also to have several such programs present in memory at any given time.

The alternative is the EXE type of program which will usually make use of more than one segment. In addition, such a program can be relocated anywhere in the memory (subject to rules about where it would start in a given segment) and will still run. This is the method that must be followed for larger programs, or programs in which data needs a separate segment. When an EXE program is loaded, MS-DOS will place it wherever it can be put.

The organization of memory in the PC type of machine is illustrated in Figure 10.1 with addresses up to 1 Mb, 1024K (in hex, 00000H to FFFFFH). On the 8088/8086 machines, this upper memory limit is determined by the number of address lines, but on the 80286 and 80386 machines the limit is determined by MS-DOS itself. The later 80486 chip machines can also make use of MS-DOS, but are primarily intended for use with OS/2 or a variant of Unix.

The normal amount of RAM memory is therefore in the 640K range, and of this DOS takes an amount which is typically 70K for versions prior to 4.0. As well as the amount taken up by DOS, it is likely that your AUTOEXEC.BAT file will see to the loading of various other programs, such as GRAPHICS, all of

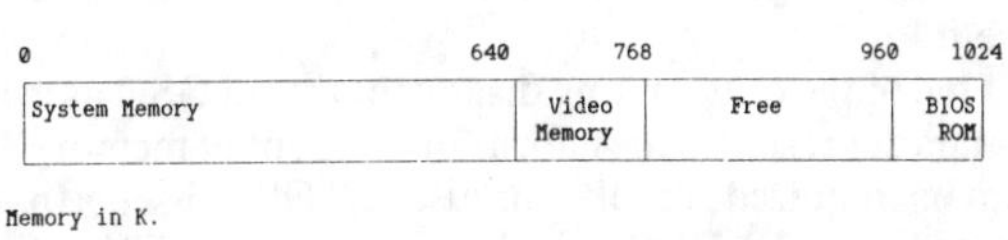

0	640	768	960	1024
System Memory	Video Memory	Free	BIOS ROM	

Memory in K.

Figure 10.1 *The organization of memory for use by MS-DOS.*

which take up more memory space. It is possible, then, to find that the lowest memory address that is free is around 140K above the start of the memory.

The OS/2 system

When the IBM PS/2 family of computers was announced, a central part of the scheme was the use of a new operating system called OS/2, which was to allow the full use of the 80286 and 80386 chips. At the time of the PS/2 launch, details of OS/2 were sketchy. It was suggested that the new operating system would be single-user and multitasking; able to utilize a full 16M of memory in an 80286 or 80386 system, and implement multitasking in any of the computers based on these chips. The operating system itself was to be used in conjunction with a 'front end' called presentation manager which would act like Microsoft Windows as a 'user friendly' front end, allowing programs to be selected from a menu by use of the mouse, and run inside windows that could be moved about on the screen. It was suggested that existing MS-DOS software could be catered for by using the bottom 1M of RAM, but that new software would have to be written specifically for OS/2. All of this was in 1987.

Since then, the progress of OS/2 has not been spectacular. The main stumbling block has been the usual one of interaction between hardware and software. At a time when no sample of OS/2 could be obtained for evaluation, no program developers could write software for OS/2, so that new software was still being written for DOS, trusting that the feature of OS/2 called the **DOS 3.X Box** would allow MS-DOS software to be run from within OS/2. With no software available that had been written or adapted for OS/2, users could hardly be expected to press for their machines to be supplied with OS/2, so that the situation developed which is still with us in 1990, and looks as if it can continue for some time, that OS/2 is available but you have to be determined and knowledgeable in order to obtain it, install it and use it.

One of the outstanding disadvantages of OS/2 as it stands at present is the enormous amount of memory that is required simply for use by OS/2 itself. In contrast to MS-DOS, which required about 70K of memory, OS/2 requires at least 1M, and will not run really satisfactorily even in this amount of memory.

A machine with 4M of memory can run OS/2 very well, but it is difficult to forecast whether this amount of memory would be sufficient to cope with the programs that will eventually be written to run under OS/2. Existing users of DOS find it difficult to justify adding RAM at a time when chip prices are high, simply in order to run programs that have yet to be written on an operating system that they have only ever seen demonstrated (if they have seen it at all). For the single user of single-tasking applications, there would seem to be absolutely no point in using OS/2 at all, and the most extensive upgrade that might be needed is to V.5.0 of MS-DOS. One point to remember here is that you can often find that in a set of apparently similar programs (such as three competitive spreadsheets) one will often work in 512K or less when the others need 1M or more. Some care in selecting software can save a large investment in hardware.

The main force that sustains OS/2 is the commitment of IBM and the desire of most users (and other manufacturers) not to be out of line. It is by no means certain that OS/2 will not be changed considerably yet again, because the system was designed around the 80286 chip and makes little or no use of the much more flexible way that the 80386 can be used. With the 80286 alleged to be a 'dead-end' system, it seems pointless to expand 80286 machines out to 4M or more of RAM, and with OS/2 unable to make really effective use of the 80386 chip, users of these machines might find a later version more suitable. It seemed inconceivable at the launch of the PS/2 range in 1987 that we would still be arguing over and waiting for OS/2 in 1990, but this looks more and more likely. The only possible advice in the circumstances is to stick with what you know and let others do the pioneering work. At the time of writing, it seemed likely that a new version of OS/2 would be able better to utilize the 80386, and would offer better support for running standard applications, probably more than one at a time.

Other DOS

Though MS-DOS, often in its PC-DOS guise, has made the running in the past this is by no means certain for the future. It was widely assumed that

MS-DOS would reach version 3.3. or 3.4 and then gradually fade away as OS/2 took over, but the launch of MS-DOS V.5.0 indicates that Microsoft does not see its OS/2 effort as necessarily being total replacement for MS-DOS. In addition, other manufacturers have new versions of older systems or totally new systems ready to challenge MS-DOS.

In particular, Digital Research, who lost out on the initial contract to provide an operating system for the first PC, now have several very useful options which are being taken up by clone manufacturers. The standard replacement DOS is DR-DOS, which offers compatibility with MS-DOS but with the useful feature of being available either in conventional disk form or as a ROM that does not require to be read from a disk. This latter form of DOS is extremely useful for portable machines which have no disk drives, or for machines in which speed of starting up is important. A machine which uses DR-DOS from ROM will have up to 604K of free RAM available if the full complement of RAM has been installed, in comparison to the 568K of RAM that is available when the operating system is read from disk into the RAM.

In addition to the set of commands which is totally compatible with the MS-DOS set, DR-DOS offers password protection for files and directories, support for networking, an optional interface for using the GEM/Desktop front end system. The latest DR-DOS also offers a full-screen text editor, a menu-driven installation program, support for hard disk partitioning of more than 32M, and support for the industry-standard system of expanded memory.

You cannot, however, necessarily decide for yourself that you will switch to DR-DOS, or that, having been provided with DR-DOS you will switch to MS-DOS, because much depends on how your computer is designed. Very often, such changes **can** be made simply by replacing all the files of one system by the files of another, but much depends on how similar the machines are. You may find, for example, that you can obtain all of the DOS commands easily enough, but that your real-time clock no longer operates correctly and you have to enter time and date each time you switch on the computer. The safest rule is to change operating systems only if you can obtain the alternative system from the manufacturer of your computer, or if you are sure that the machine is truly IBM compatible.

Concurrent DOS XM, also from Digital Research, is aimed at the 8086 and 80286 machines, and allows low-cost terminals to be fitted to a computer to allow for up to five users who can share the files, printer(s) and disks. The usual range of applications that are normally run under MS-DOS will run under Concurrent DOS XM in this form, which costs considerably less than networking, though each application that is run in this way slows down the effective computing speed, though this is not particularly noticeable unless all of the users are trying to run spreadsheets at the same time.

Concurrent DOS XM also offers multitasking capabilities, so that you can run up to four applications programs in separate windows on the screen of the main computer, moving from window to window. This is a built-in part of the operating system itself rather than an add-on, and to take advantage of it, Concurrent DOS XM can make use of expanded memory boards holding up to 8M of RAM which will be used by the programs. This is a useful point – many systems that claim to make use of expanded RAM use the RAM only as a disk cache or as RAMdrive, not as memory in which programs can be run.

Another operating system from Digital Research is Concurrent DOS 386 which, as the name suggests, allows the more advanced features of the 80386 chip to be used. Up to ten users can be accommodated, sharing files, disks and printers, and with each extra user working on a simple low-cost terminal. Multitasking is also permitted, using up to four windows on the main computer, up to two applications for each user on a terminal. The full memory capabilities of the 386 chip can be used, with support for up to 4G of memory (1G=1024M), and the memory can be organized into banks without any additional software. If applications require that extra memory is to the LIM standard, the operating software can simulate this.

All of this convenience comes at a price, and Concurrent DOS 386 is quoted at £395, which is small enough if you need the multiuser capabilities. Installation is comparatively easy, but you should check before taking the plunge that your computer is suitable – in general any computer that is 100 per cent IBM compatible should present no problems in installing Concurrent DOS 386.

Unix

Unix is an operating system that is of older origin than MS-DOS or even CP/M, and its potential advantage is that it could cope with the full addressing range of the 80386 chip, with multiuser and multitasking built in rather than added on. The disadvantage is that it is totally incompatible with MS-DOS so that existing applications software could not be run unless your version of Unix incorporates an MS-DOS emulator.

Unix originated in 1970 as an operating system for minicomputers and it was by 1972 rewritten in the C programming language. The original (source-code) text for Unix was then made available at very low cost (or even given away) to anyone who could use it, and anyone who obtained a copy was encouraged to use it and to modify it in any way that seemed appropriate. This resulted in many incompatible versions of an operating system that should, on any common-sense basis, have been standardized. As far as computers of the PC class are concerned, Unix is now available in two versions of which one, the Open Desktop from SCO (Santa Cruz Operation) offers MS-DOS emulation. The Open Desktop package also overcomes the main objections that commercial users have always voiced about Unix, that it was a system that relied on a set of incomprehensible commands, by using a graphical front end so that the user simply picks a program or action icon by using the mouse, rather in the style of Microsoft Windows and other such packages.

Some observers expect Unix to be the major competitor to OS/2 as the operating system for 80386 and 80486 machines in the 1990s. This is probably expecting too much of the human capacity to cooperate, and it is just as likely that we shall see an MS-DOS 6.0 or a new DR-DOS to make better use of the 386 and 486 chips without requiring so much memory as to make a 1M machine look starved of RAM. The crucial factor is to what extent users will feel that all the windowing, icon moving and mouse action is worth the huge amount of memory that it uses when by dispensing with such things they could run well-established programs quickly and easily. Certainly since the first Apple Macintosh appeared, it has been taken as the conventional wisdom that all future computers will use icons, mice and windows. If, however, you have seen demonstrators fumbling with a mouse for a time that would have allowed you to type DIR and (RETURN) some ten times, you might question this conventional wisdom and argue that since virtually all of your useful data input to the machine is by way of the keyboard, it seems silly to abandon that path for the relatively small number of DOS commands. Perhaps in the 1990s some reasonable compromise will be found – but computing is not renowned for compromises.

MS-DOS interactions

The use of MS-DOS implies that computing actions can be carried out by loading registers and then forcing a software interrupt which uses a specific number. The most important of MS-DOS interactions are with assembly language, but many high-level languages make provision for direct calls to MS-DOS.

The assembly language form of a numbered interrupt is INT **n**, where **n** is a number in the range 0 to 255 inclusive. When this is translated into machine code it appears, in hex, as the number code **CD** followed by the value of **n** in hex, so that CD 20 means interrupt number 32. It is more usual to keep to hex numbering for these interrupts, however.

The possible range of interrupts contains many which are reserved for calling in special circumstances, and which relate to the microprocessor rather than to the use of MS-DOS. The first set of interrupts, for example, with all numbers in hex, is as shown below, though the use of interrupts 5 to 1F will vary from one machine to another:

INT 00 divide by zero error
INT 01 single step (used by DEBUG T command)
INT 02 non-maskable interrupt
INT 03 breakpoint interrupt (used by DEBUG)

INT 04	overflow in multiplication
INT 05	print screen (software interrupt)
INT 06	mouse button control (software interrupt)
INT 07	reserved software interrupt
INT 08	system clock
INT 09	keyboard
INT 0A	real time clock
INT 0B	communications 1
INT 0C	communications 2
INT 0D	hard disk
INT 0E	floppy disk
INT 0F	printer
INT 10	VDU (software interrupt)
INT 11	system configuration (software interrupt)
INT 12	memory size (software interrupt)
INT 13	disk input/output (software interrupt)
INT 14	serial input/output (software interrupt)
INT 15	reserved for enhancement (software interrupt)
INT 16	keyboard input/output (software interrupt)
INT 17	printer input/output (software interrupt)
INT 18	system restart (software interrupt)
INT 19	disk bootstrap (software interrupt)
INT 1A	system clock and real-time clock in/out (software interrupt)
INT 1B	keyboard break (software interrupt)
INT 1C	external timer interrupt (software interrupt)
INT 1D	initialize VDU parameter table (software interrupt)
INT 1E	disk parameter table (software interrupt)
INT 1F	matrix table for VDU (software interrupt)

INT numbers 14–1F are reserved for use by future Intel chips, so that the use of these interrupts would make programs potentially incompatible. There are also two unofficial interrupt numbers, 28H and 29H.

MS-DOS interrupts

MS-DOS can make use of the interrupts 20H to 3FH which are reserved for MS-DOS use with the Intel chip set. Of these, the INT 21H is by far the most important, as this is the way in which virtually all of the common MS-DOS actions are called up. The use of INT 21 will therefore be dealt with in much more detail than the others which are noted in the table. Of these, some are reserved for future expansion, and it is important if programs are being written for a variety of machines that the set of actions which is used should be common to at least the versions of MS-DOS from 3.0 onwards. Though at the time of writing, MS-DOS has reached V5.0, the differences between 5.0 and 3.0 relate mainly to actions that will not concern the majority of programmers who are concerned with adapting programs or with writing short program sections. The maintenance of compatibility is more important than the ability to use advanced features, particularly since there will be machines using versions 3.0 to 3.2 for a considerable time to come. It is only comparatively recently that it has been possible to start ignoring the earlier versions 1.x and 2.x in writing new programs.

MS-DOS interrupts in detail

The MS-DOS interrupts contain several duplications of actions, many of them due to the evolution of MS-DOS over the years. In the following table, the interrupts are described only in outline, and a full listing of the INT 21H set follows. The interrupts described as **unofficial** are not documented by MS-DOS but have been exposed by others and reported on. Their use is discouraged, since such uses could change from one version of DOS to another.

INT 20H	End of program. Seldom used now, because there is a better version in INT 21H.
INT 21H	The MS-DOS function interrupt which allow the use of the functions listed later.
INT 22H	End of program vector, the address to which the machine goes when a program ends.
INT 23H	Ctrl-Break address for the start of the routine that takes over when Ctrl-Break is pressed.
INT 24H	Critical error address for the start of the routine that takes over when a critical error

	occurs. These are: write-protected disk, unknown unit, drive not ready, unknown command, data error, bad request structure length, seek error, unknown media, sector not found, printer out of paper, write fault, read fault, general failure.
INT 25H	Disk read of specified track/sectors.
INT 26H	Disk write of specified track/sectors.
INT 27H	Terminate and stay resident (older method).
INT 28H	Unofficial re-entry to DOS for TSR programs.
INT 29H	Unofficial character output.
INT 2AH–2EH	Reserved for future use.
INT 2FH	Time-sharing interrupt, used by PRINT.
INT 30H–3FH	Reserved for future use.

The INT 21 functions

A call to INT 21 is used to carry out an action of MS-DOS, and the call involves three steps:

1. A function number is placed in the AH register.
2. Other numbers, as required by the action, are placed in other registers of the microprocessor.
3. The INT 21 is executed.

Each of the functions is documented in such a way that the contents of the registers before and after the call can be determined. If registers are likely to be modified by a call, the programmer may need to save these registers on the stack in the usual way. The use of INT 21 is therefore very extensive, since as many functions can be catered for as there can be numbers stored in the AH register. The following list is in numerical order for ease of looking up rather than in the order of probable use, and describes the functions briefly. For a a full description, consult a text of MS-DOS programming. Note that many functions relate to the file control block system which was used for file identification in the 1.x versions of MS-DOS, and are retained only for compatibility purposes. Use the system of file handles (introduced with V.2.0) for all programs being written, unless there is a particular need to maintain compatibility with Versions prior to 2.0.

When a function is called, it may cause an error, and there has to be a provision for reporting the error. The method that is used is to set the carry flag if an error has occurred (clear otherwise) and to return an error number in the AX register. The error codes are noted later in this section. Functions 00H to 2EH are used in all versions of MS-DOS. Functions 2FH to 57H are used in version 2.0 onwards, and functions 58H onwards are used in V.3.0 onwards, though function 63H is used in V.2.25 and in no other version. V.3.3 functions start at 65H, and functions from 69H start to appear on V.4.0.

A few descriptions start with the word **unofficially**, meaning that the use is undocumented by Microsoft, and may be changed. Such functions should be avoided, and if they are found in software, they might need to be replaced.

Function 00H – used to terminate COM programs only
Function 01H – read key, echo on screen. Character code is returned in AL register.
Function 02H – byte in DL register displayed on screen.
Function 03H – serial port input, byte in AL register.
Function 04H – serial port output, byte in DL register.
Function 05H – parallel printer output, byte in DL register.
Function 06H – keyboard input or screen output depending on value in DL register.
Function 07H – key input, no echo, character in AL register.
Function 08H – as 07H, but checks for Ctrl-Break.
Function 09H – print string whose address is in DS:DX. String end marked with $ which is not printed.
Function 0AH – key input, buffered. DS:DX set to start of buffer, of which first byte is a length byte, second byte will be set to number of characters, and string starts at third byte. Carriage return ends input, 'bell' sounds if buffer is filled. Editing keys can be used on buffered characters.
Function 0BH – check input. AL register contains

FFH if any input character available, 00H otherwise.
Function 0CH – clear buffer and read. AL register must contain a read code, such as 01H. The keyboard buffer will be cleared, and a read carried out.
Function 0DH – clear disk buffers.
Function 0EH – drive select, using number in DL register. After use, AL contains number of drives (minimum 2 even on a single-drive machine).
Function 0FH – file open using file-control block (FCB) system (for compatibility with older versions). DS: DX set to address of the FCB, which must not already be open.
Function 10H – file close, update director. DS: DX contains address of open FCB.
Function 11H – match filename, using FCB. DS: DX contains address of FCB, and the directory will be searched for the name in this FCB, which can contain wildcards. AL contains 00H if successful.
Function 12H – next match search, will continue the search started by function 11H.
Function 13H – file delete using FCB. DS: DX contains address of FCB.
Function 14H – read sequential file using FCB. DS: DX contains address of open FCB, data is read into disk transfer address (see Function 1AH).
Function 15H – write sequential file using FCB. DS: DX contains address of open FCB, which contains details of file to be written.
Function 16H – file create using FCB. DS: DX contains address of unopened FCB. Function checks directory for existing file, or free space.
Function 17H – file rename using FCB. DS: DX contains address of FCB.
Function 18H – reserved.
Function 19H – default drive number returned in AL register.
Function 1AH – set address for disk transfer in DS: DX. If this function is not used, the address CS: 0080H will be used.
Function 1BH – disk FAT (partial) information. After use, DS: BX gives address of start of FAT of default drive in memory, DX contains number of allocation units, AL sectors per byte, CX bytes per sector. Applicable mainly to V.1.x.
Function 1CH – disk FAT as 1BH, but for specified drive rather than default drive.
Functions 1DH–20H – reserved for future use.

Unofficially 1FH gets the drive parameter table for the default drive.
Function 21H – random access read using FCB. DS: DX contains address of open FCB, record read into disk transfer address.
Function 22H – random access write using FCB. DS: DX contains address of open FCB, record written from disk transfer address.
Function 23H – file size using FCB. DS: DX contains address of unopened FCB, whose record size values will be altered if the file is found.
Function 24H – random access record number using FCB. DS: DX contains address of open FCB, whose relative record field number will be altered.
Function 25H – interrupt vector set. DS: DX contains address of first byte in interrupt table, and AL contains interrupt reference number.
Function 26H – create new program segment – V.1.x only.
Function 27H – random access block read using FCB. DS: DX contains address of open FCB, CX contains number of records whose size is specified in the FCB. Records are read into the disk transfer address.
Function 28H – random access block write using FCB. DS: DX contains address of open FCB, CX contains number of records to be written from the disk transfer address.
Function 29H – parse command line tail for FCB. AL contains status byte to determine how separator characters, drive numbers, filenames and extensions are treated. After use, ES: DI contains address for FCB, DS: DI contains address for first character following command line.
Function 2AH – Date returned in CX (year), DH (month number), DL (day number in month), and AL (day number in week).
Function 2BH – set date, using CX, DH and DL as above. On Amstrad and other machines with battery-backed clock, 2BH resets the date on this clock.
Function 2CH – time returned in CH (hours), CL (minutes), DH (seconds) and DL (hundredths of seconds).
Function 2DH – set time, using CH, CL, DH, DL as above. Battery-backed clocks will be reset.
Function 2EH – disk file verification set/reset. AL contains 00H to turn verify off, 01H to turn verify on.

The following functions are not used in V.1.x of MS-DOS.

Function 2FH – find disk transfer address, place in ES: BX.
Function 30H – find DOS version number, place into AL with figure after decimal point in AH. AL contains 00H if a version prior to 2.0 is being used.
Function 31H – terminate program and stay resident (TSR). AL contains exit code, DX stores number of 16-byte blocks that must remain allocated to the resident program.
Function 32H – **unofficially** used to give address of drive parameter table in DS: BX.
Function 33H – Ctrl-Break flag. IF AL set to 00H, Ctrl-Break setting is reported only. With AL=01H, DL=0 turns Ctrl-Break off, DL=1 turns Ctrl-Break on so that it will be checked at each call to DOS rather than on input/output only.
Function 34H – **unofficially** returns in ES: BX address of counter which records when DOS has been entered (so as to prevent re-entry).
Function 35H – find interrupt address for INT number in AL. Address is returned in ES: BX.
Function 36H – find free space on disk, drive number in DL. After use, BX contains number of free clusters, DX total number of clusters on drive, CX number of bytes per sector, AX number of sectors per cluster.
Function 37H – **unofficially** find or reset the command line option character, which is usually /.
Function 38H – find or reset country information, V.2.x can find only. For V.3.x onwards, AL is loaded with FFH, BX contains country number and DS: DX holds address for a buffer which will contain the data (34 bytes max.).
Function 39H – create subdirectory. DS: DX contains address for path name of new subdirectory, using 0H terminator.
Function 3AH – remove subdirectory. DS: DX contains address for path name of subdirectory, terminated with 0H. Subdirectory must be empty, and must not be in current use.
Function 3BH – change current subdirectory. DS: DX contains address for path name of subdirectory, terminated with 0H.

Functions 3CH to 46H were introduced in V.2.0 as a

way of making file actions simpler and faster, replacing the file control block method that was inherited from CP/M with an improved method called file handles. These functions should be used in any programs being written for file actions. When a file is created, using function 3CH, or opened using function 3DH, a 16-bit number called the **file handle** is placed in the AX register, and this number is from then on used as a reference to the file. In V.2.0 onwards, it becomes possible to write to or read from peripherals by using names such as PRN, CON, AUX1, etc. This is made possible by assigning reserved file handle number to these peripherals as follows:

00H	Keyboard input
01H	Screen output
02H	Error report
03H	Auxiliary
04H	Printer

Function 3CH – file create using handle. DS:DX contains address for string containing drive, path and filename, terminated with 0H. Attribute code is placed in CX. After use, AX contains file handle if carry flag is reset, otherwise AX carries error code.
Function 3DH – file open using handle. DS:DX contains address for string containing drive, path and filename, terminated with 0H. AL contains access code for read and write by owner or other network users. Use only codes 0H (write), 01H (read) or 02H (read and write) if the machine is not to be networked. After use, AX contains file handle or (if carry flag set) an error code.
Function 3EH – file close using handle. BX is used to hold file handle number.
Function 3FH – read file or device using handle. BX is used to hold file handle number, CX the number of bytes to read, and DS:DX contains the address for a buffer in memory. After use, AX contains the number of bytes read or (if carry flag set) an error number.
Function 40H – write to file or device using handle. The file handle is placed in BX, the number of bytes to be written in CX, and the buffer address in DS:DX. After use AX contains the number or bytes written or (if carry flag set) an error code.
Function 41H – delete directory entry. The DS:DX

registers are used to contain the address of a string containing drive, path and filename, terminated with 0H.

Function 42H – file pointer move. When a file is created or opened, the first byte is available for reading or writing, and the byte position is held as a pointer number. By using this function, the 'pointer' can be moved so as to read any byte out of order. The pointer number is held in CX: DC, the file handle in BX and AL is used to hold a code that determines whether the pointer number is added to the start of the file number, end of the file number or current pointer number. After use, DX:AX stores new pointer number, but if carry flag is set, then AX contains an error code.

Function 43H – File attribute change. DS: DX contains address for a filename string containing drive, path and name, terminated with 0H. AL is set to 00H to report attributes, 01H to alter attributes. CX is used to hold the new set of attribute bits if these are to be changed.

Function 44H – input/output control. File handle is put into BX, using numbers 00H to 04H (see list above) for devices. Sixteen different forms of the function can be called by using values 00H to 0FH in AL. These are listed in outline here, with no details about resister contents.

AL Value	Action
00H	get device channel 16-bit data in DX
01H	set device channel 16-bit data in DX
02H	read string from device.
03H	write string to a device.
04H	read string from disk.
05H	write string to disk.
06H	get file or device input status.
07H	get file or device output status.
08H	check for fixed disk.
09H	test is drive local or remote on network.
0AH	test file handle for local or remote.
0BH	change retry count for file sharing in network.
0CH	change device code page.
0DH	multi-action function in V3.2 onwards – used to get parameters, set

	parameters, format, read track or write track on disk.
0EH	get last logical drive letter.
0FH	assign logical device letter.

Function 45H – file handle duplication. BX contains a file handle, and after the action, AX contains a handle number for the same file.
Function 46H – file handle forced duplication. BX contains one file handle and CX another. After use, both handles refer to the same file, and if the CX handle referred to another open file, that file is closed.
Function 47H – get current directory. DS:SI contains address of a memory buffer, DL contains a drive number. After use, the buffer contains the string with directory path (no drive letter) terminated with 00H.
Function 48H – memory allocation. BX is used to specify number of 16-byte units (paragraphs) to be allocated. After use, AX contains segment address.
Function 49H – memory release. ES contains segment address of memory block which is to be returned for reuse.
Function 4AH – memory reallocation. ES contains segment address of allocated block, BX contains number of 16-byte paragraphs in modified block.
Function 4BH – child load and execute. Allows a program that is running (the **parent**) to load and run another program (the **child**). Memory must have been de-allocated using function 4BH, because MS-DOS normally assigns the whole of available memory to a program. Used by applications programs to allow use of MS-DOS commands by making COMMAND.COM a child program. AL contains code determining if child is to be loaded and run, or overlaid at a specified address and not run. DS:DX contains address for string of drive, path and filename for child program, terminated with 00H. EX:BX contains address of memory that contains any information needed by the child program (the environment parameters).
Function 4CH – terminates a program. Returns a code in AL which can be picked up as an ERRORLEVEL code in a batch file (or for a child program with Function 4DH).
Function 4DH – get child program return code. The

AL register contains the return codes:

00H	Normal termination
01H	Ctrl-Break used
02H	Error caused termination
03H	TSR

Function 4EH – search for matching filename. DS: DX contains address of string containing drive, path and filename (wildcards permitted), using 00H as terminator. CX can contain attribute bits. If a matching file is found in the directory, file details will be loaded into the disk transfer area.
Function 4FH – find next matching file. Continues action that was set up by use of Function 4EH.
Function 50H – **unofficially** designates a program to be current if its ID number is loaded into BX before calling 50H.
Function 51H – **unofficially** gets the ID number for a current program in BX register.
Function 52H – **unofficially** gets address in ES: BX for a table of values held in memory.
Function 53H – **unofficially** gets into DS: SI address for a BIOS parameter block, with ES: BP containing address for drive parameter table.
Function 54H – verify state. After use, AL=00H if verify is off, AL=01H if verify is on.
Function 55H – reserved for later use.
Function 56H – file rename. DS: DX is used to hold the address of the string that contains drive, path and filename, ending with 00H, and ES: DI contains the new string (using the **same** drive).
Function 57H – read/write time and date on file. BX must contain the file handle. If AL=00H, then after use file date is in DX, time in CX. If AL=01H, then the values in DX and CX will be written to the file.
Function 58H – memory allocation method. When memory is to be allocated, MS-DOS can allocate by searching either from the start of memory or from the end, and by allocating any block that is large enough, or looking for the block that is closest to the required size. AL is used to hold the get (00H) or set (01H) code, and AX is used to return the current strategy code, or BX to contain the code that is to be used.
Function 59H – extended error messages. Must be used immediately following an error with BX set to 00H. After use the error code is in AX, with an error

class code in BH, advised action in BL and error source (unknown, block device, network, serial device or memory) code in CH.
Function 5AH – create temporary (scratchpad) file. DS: DX contains address for string containing path, ending with backslash and zero. After use, DS: DX will contain specifier string for the temporary file, ending with 00H.
Function 5BH – create new file (compare 3CH). DS: DX contains address for specifier string, CX contains attribute code bits. After use, AX contains file handle, **but** not if the file already exists.
Function 5CH – file lock/unlock for networking. File handle is placed in BX, and AL holds 00H is file is to be locked (01H to unlock). Registers CX: DX contain the address for the start of the region to lock, with SI: DI holding the length of the region.
Function 5DH – **unofficially** used to obtain information on some types of errors.
Function 5EH – used for networking with 5EH in AH and codes 00H – 03H in AL.

AL=00H	get machine name
AL=02H	send set-up string to printer
AL=03H	show set-up string for printer

Function 5F – used for networking with 5F in AH and codes 02H – 04 in AL.

AL=02H	get list linking device name with network name.
AL=03H	alter redirection of device to network name
AL=04H	cancel redirection

Functions 60H and 61H not currently used.
Function 62H – find PSP. The program segment prefix is the table in memory which occupies bytes 0000H to 00FFH (so that the program starts at 0100H in the segment). This table contains data about the program such as terminate address, exit address for Ctrl-Break and so on. Use 62H in AH to call, and after use BX has segment address of the PSP.
Function 63H – used in V.2.25 only, do not make use of this function.
Function 64H – not currently used.
Function 65H – get code page.
Function 66H – set code page.

Function 67H – set number of permitted file handles. The default is 20, and to change this number, call the function with BX containing the new number of handles. On V.3.x, avoid using even numbers or numbers close to FFFFH.
Function 68H – clear buffers. BX contains a file handle and when the function is used, all buffers associated with the file are cleared and the directory is updated.

Error codes
When an error occurs during an MS-DOS function, the carry flag will be set, because the state of this flag is the easiest to test. When this happens, an error code will be returned, usually in AX, and many of the codes are standardized, though some functions return error codes of their own. The standard error codes are shown below, with codes 01H to 12H being used on all versions of MS-DS, codes 20H onwards on versions 3.0 onwards.

Code	Meaning
01H	function number not valid
02H	file not found
03H	path not found
04H	no spare handles (too many open files)
05H	access denied
06H	handle not valid
07H	control block in memory corrupted
08H	out of memory
09H	memory block address not valid
0AH	environment not valid
0BH	format not valid
0CH	access code not valid
0DH	data not valid
0EH	reserved
0FH	drive specification not valid
10H	cannot delete current subdirectory
11H	not same device
12H	no more files

The following error codes apply to network use and will not be found when a single-user machine is involved.

20H	sharing violation
21H	lock violation

22H	disk change not valid
23H	FCB not available
24H	sharing buffer overflow
25H–31H	reserved
32H	network request denied
33H	remote computer not receiving
34H	name duplicated on network
35H	network name not found
36H	network busy
37H	device does not exist on network
38H	BIOS command limit on network exceeded
39H	hardware error on network adapter
3AH	network response incorrect
3BH	network error (unknown)
3CH	remote adapter incompatible
3DH	print queue full
3EH	print queue not full
3FH	print file deleted
40H	network name deleted
41H	access denied
42H	device type incorrect
43H	network name not found
44H	network name limit exceeded
45H	exceeded network BIOS session limit
46H	temporary pause
47H	network request denied
48H	redirection temporarily halted
49H–4FH	reserved
50H	file already exists
51H	reserved
52H	cannot create directory entry
53H	INT 24 failure
54H	too many redirections
55H	duplicate redirection
56H	password not valid
57H	parameter not valid
58H	fault in network device

11 Programming techniques

The Intel chips can be programmed either in their native machine-code or by way of operating system calls as noted in the previous section. Since this book is concerned mainly with the hardware of the Intel processors, programming by way of a high-level language does not concern us, and this section is concerned only with an outline of machine-coding methods. For a fuller treatment of Intel chip programming, the reader is referred to any of the large number of texts that are entirely devoted to the subject.

The normal method of programming is to make use of code developed by way of assembly language on a PC-compatible computer or on another type of machine using a cross-assembler. Assembled code, after extensive testing, can be contained in ROM or on disk for use with the chip system. Assembler programs are available from a number of suppliers, notably Microsoft, and they generally follow the Intel guidelines with minor deviations. One very fast and capable assembler program, called A86, is available on a shareware basis and is extensively used by machine-code writers. A86 was written by Eric Isaacson, formerly a senior engineer at Intel, and is strongly recommended to anyone who needs to program the 8086/8088 chips or to develop programs for the later chips.

The action of the assembler is to convert statement lines written in assembly language into suitable machine code. The statement lines are created as an ASCII file from any text-editor or word-processor that can produce such files – note that all word-processors can produce ASCII files but few do so as a default, whereas text-editors will produce **only** ASCII files. The assembly action is invoked on the computer by typing the name of the assembler program followed by a space and then by the name of the ASCII file of assembly language. In addition, various letters may be added, usually separated by slashmarks, in order to compel the assembler to work in some particular way such as using hexadecimal numbers as default, omitting FWAIT instructions for use with 80287, taking account of the case (upper or lower) of letters and so on. These additions are called assembler switch settings, and for routine use, programmers will create a batch file into which the preferred settings can be incorporated to be used each time assembly is required.

Programs can be written so as to assemble into a COM type of file, which will be a program that is ready to run, using one single segment for code, data and stack, or to OBJ modules. If OBJ modules are specified, the type of code that is generated is intermediate, and cannot be used until the modules are linked together. This latter method is used to create the EXE type of file, and is also used when a program is written as a set of self-contained modules, or when an assembly-language module has to be attached to modules that have been written in high-level language.

The assembler deals with three types of statement lines, instruction statements, data allocation statements and assembler directive statements. The instruction statements make use of the standard Intel mnemonics such as MOV, PUSH, OR, JMP along with whatever operands are needed. Numbers can be included in immediate form, as register references (number contained in a register) or as memory references. For all but the shortest programs, memory addresses will be used in the form of **label names** rather than in the form of numbers. All assemblers allow for a statement to be preceded by a label name which will then be taken as a reference to the address at which the statement will be assembled. For example, the statement:

```
MOV AX, datlist
```

will copy into the AX register the address corresponding to the label **datlist**. This label would be placed at the start of a set of statements which established data values. Similarly, the statement:

```
JMP newone
```

would cause a jump to a section of code whose

starting point was marked by the use of the label **newone**, as, for example, by:

NEWONE: MOV AX,DATA

The use of labels for all memory references and data numbers is virtually essential if a program is to be comprehensible, and the values of all labels can be printed out by the assembler program in the form of a **symbol-table**. In addition, the programmer should use comments (any line or end of line marked with a semicolon) to indicate what use is being made of labels to indicate numbers. Comments are not neccesary for labels which are so frequently used as to be 'standard items', like CR to mean 0DH, the carriage return character, but should show the reason for using other labels which are likely to be specific to a program.

Label names that precede a piece of code need to be separated by the use of ı2arated by the use of a colon, but where abels are used to define the address of data (in the data segment) no colon is needed. Some assemblers insist that each statement is divided into separate fields (using the Tab key to separate fields), others allow statements to be entered with only a space separating each portion of the statement.

Data allocation statements are used to assign data numbers to an address or list of addresses. Such statements will be used to place values into tables or to point to strings, and each data allocation makes use of a two-letter code which specifies the type of data. The processors can work with the fundamental data types of byte (8-bits), word (16-bits) and Dword (32-bits), and the floating-point co-processors can manipulate 8-byte quantities, referred to as Qwords and 10-byte quantities referred to as Twords. The codes for these data types are DB, DW, DD, DQ and DT respectively, so that:

size DW 4F2CH

means that the word **size** is used as the address at which the word 4F2CH is stored. The storage will be low-byte first, so that 2CH is stored at the address corresponding to **size** and the byte 4FH is stored at address **size + 1**. If no value is specified, the memory space is allocated but unassigned, and this can be marked by using the **?** character, so that:

count DW?

will reserve memory space for a word starting at label **count**, but without assigning a value – by convention, the **?** mark would assign zeros into the memory of these two bytes.

A value or set of values following a data allocation code can be used both to assign values and to allocate spaces. For example, a string can be assigned by using a statement such as:

explain DB "False entry"

in which the statement assigns 11 bytes and fills them with the ASCII codes for the phrase, marking the start of this block with the label. Another form of assignment is provided by the DUP statement, which allows a specified number of memory locations to be filed with the same value. For example:

temp db 100H DUP 0FH

will fill 100H bytes of memory with the data 0FH, marking the start of this block with the label **temp**. Constructions like DUP(?) can be used to ensure that space is not initialized to any particular values.

The assembler directive statements consist of instructions which are more likely to be specific to one particular assembler program. In this respect, some assemblers are very much easier to use than others, and the terminology that is used by some assemblers can be very confusing. The examples here are from A86, which is by far the most rational in this respect, and are only a small selection of directive which are likely to be found on other assemblers.

Assembler directives are concerned with allocation of memory and with segmentation. An assembler can support the use of multiple segments but in practice programs that will take up more than one 64K segment are always written using higher-level languages, nowadays usually C. Assembly language, which is very much more difficult to write in large chunks, is confined to shorter programs whose code, data and stack can be held in one single segment.

Assembler directives such as CODE SEGMENT, DATA SEGMENT are therefore used only to mark out differences to the reader of the program, in the sense that the code in the code segment is generated by the assembler program and the code in the data segment is provided by the user or generated when the program runs. Assemblers provide for either COM or OBJ output, and the default should be COM, implying that a single segment will be used for code, data and stack. The main use of OBJ code is to allow a new module to be developed to add in to an existing program by relinking the modules of that program. This is a very common method of linking a piece of assembled machine code into an existing program written in a high-level language.

A directive that is found in all assemblers is EQU, which allows a variable name to be used for a number. For example:

```
MAX EQU 2FFH
```

will mean that the name MAX can be used in place of the specific number 2FFH. The advantage of using variable names is that where a particular number is used extensively in a program, a change of number can be carried out by altering one line (then reassembling) rather than by replacing each occurrence of the number. Note the difference between a label, which is the **address** of data, and a variable name, which **is** the item of data.

The ORG directive allows for a specified memory address to be used, usually for placing data. For example:

```
DATA SEGMENT
ORG 03500H
```

will have the result of starting the data at address 03500H – despite the use of the word SEGMENT this need not imply that a separate data segment is used, only that the data will be contained in a block starting at this address. Most assemblers require no ORG statement for code, and will assemble code starting at offset address 0100H in the code segment.

EVEN allows for data to be put into memory starting at an even memory address. This is particularly useful for the 8086 and later chips which can perform faster 16-bit read and write actions when data is aligned so as to start on even-numbered addresses. There is no speed advantage to be gained on the 8088/80188 type of chip.

PUBLIC is used to define a list of symbols whose values will be defined in the current module and which can be used by other OBJ modules when they are linked together.

EXTERN is used to mark symbols which are not defined in the current module but which are to be defined in some other module. This allows the assembler to allow correctly for the use of the symbol, so that the correct translation will be inserted when the modules are linked.

MAIN is used in a modular program to identify the module which contains the starting address for the entire program, so that only one module of a set should contain this directive.

Macros

Assemblers for the Intel chips generally incorporate a macro facility so that statements that have to be repeated many times over can be represented by a macro word which can be followed by register, label or variable names. Beginners often confuse a macro with a subroutine (or procedure). A subroutine is a piece of code which can be entered from several different points in the code (using CALL) and which returns to the instruction following the CALL instruction. A macro placed into an assembly generates a piece of code at the point at which it is placed, so that if a macro is put in ten times, it will generate the same piece of code in ten places rather than generating one piece of code that is called from ten places.

The form of a macro is Name MACRO Variables, starting with the name that will identify the macro, then the directive word MACRO, and followed by a list of the registers, labels or variables names that the macro will use. This latter list is given in forms such as arg1, arg2, arg3 or #1, #2, #3, which show the order in which these items will be dealt with. When the macro is used, the list of these **arguments** will provide the correct set of names. The end of the macro lines will be marked by ENDM or #EM, depending on the assembler you are using.

For example, a macro to add two numbers might appear as:

```
SUMMIT MACRO #1,#2
MOV AX,#1
ADD AX,#2
#EM
```

which will load the first number into AX, then add in

the second number so that the sum is contained in AX. A macro like this could be put into an assembly language listing in the form:

SUMMIT NUM1, NUM2

which will take the numbers represented by NUM1 and NUM2 and provide the sum in the AX register. The code that appears in the listing when this macro has been used is the code for the lines:

```
MOV AX,NUM1
ADD AX,NUM2
```

Working with data

Note: Assemblers will indicate as an error any attempt to load a single-byte register with a word, or a word register from a byte register.

Load and Store

These actions use the MOV instruction. MOV takes the syntax:

MOV destination, source

and can use memory reference, register, or immediate data, though it cannot be used to move data from one memory address to another directly. A label-name cannot be used as a destination for a MOV instruction. MOV is used in its simplest form for data, but can be used for addresses when the word OFFSET is added to the name of the data. For example:

MOV CX, OFFSET WRDDAT

will load into CX the address of the data called WRDDAT.

This latter type of action is more easiy carried out using LEA, with the syntax:

LEA CX, WRDDAT

which can also locate the address of a table value using:

LEA CX, WRDDAT[SI]

where the SI register carries the index number in the table.

The LDS and LES instructions will result in four bytes being loaded, the lower two into a specified register and the upper two into DS or ES. For example:

LDS BX, ADDR

will load from ADDR and ADDR+1, and load DS from ADDR+2 and ADDR+3

Arithmetic
The ADD/ADC and SUB/SBB actions use the same destination, source syntax as MOV and are used for integer addition and subtraction. The use of the variants ADC and SBB allows for a carry or borrow to or from a previous action, so that multibyte arithmetic is easy to program for.

The use of multiplication and division is also available, using either signed or unsigned methods for bytes or words. In byte actions, one byte is held in AL and the other in memory or in another 8-bit register. The result of the multiplication or division is then held in AX. For example:

MUL BL

will multiply the number in BL by the number in AL and hold the result in AX.

When word quantities are multiplied or divided, one quantity is held in AX and the other in memory or in another 16-bit register. The result is then held in DX:AX as a Dword.

Logic actions
The AND, OR and XOR actions take the usual destination, source syntax, and can use the usual range of addressing methods. The AND action is used predominantly for masking bits in a byte or word, as for example:

AND BL, OFH

which will ensure that the byte in BL contains only zeros in its upper nibble. The OR action can be used to ensure that specified bits are set to 1, so that:

OR AL, 80H

will ensure that the most significant bit in AL is set. The XOR command can be used to clear a register, using the form:

XOR AL, AL

or to detect that a bit or word has changed in comparison to a standard.

The logic set also includes the NOT action which will complement (invert) a word or byte.

Shift and rotate instructions can be used to shift by one place or by the number of places specified by a number stored in the CL register. Typical forms of command are:

SAL AX,1 ; shift word in AX by one place left
ROL addr1,CL ; rotate to the left the word at address using number of places specified in CL.

Branching

Unconditional branching uses JMP, and when this is followed by a label name, the assembler requires this label to be defined as **near** (in the same segment) or **far** (in another segment) unless assembly is to code in a single segment to make a file of the COM type. If the jump is designated as SHORT, it is restricted to ±127 bytes from the jump point. JMP is a 3-byte instruction, and JMP SHORT needs only two bytes.

One form of indirect jump is provided by using a 16-bit register, in the form:

JMP BX

where the register contains the address within the segment. The other indirect jumps use a register to point to a word in memory which provides the address, and there is provision for using a table of such pointer words.

For long jumps (to another segment), the JMP label method can be used with the label name declared as FAR. An indirect jump using Dword can also be used to branch to another segment, so that:

JMP WORD PTR [BX]

will use the content of BX to find a word in memory that in turn provides the jump destination, and:

JMP DWORD PTR [BX]

will use the content of BX to find a Dword (4 bytes) in memory that provide both segment number and offset (loading CS and IP) for the jump destination. The 80286 and 80386 also provide jumps to task switches.

Conditional branching makes use of flags, and these flags may have been set by arithmetic actions or by the use of CMP or TEST. The CMP instruction performs a subtraction so as to set flags, but does not alter the contents of any register or memory. The TEST instruction performs a logical AND, once again setting flags but not altering registers or memory. For example:

CMP memplace, CX

will subtract the contents of CX from the contents of **memplace** (both words) and set flags accordingly, but without altering either content.

The conditional branching instructions each provide a branch for a particular flag setting or combination of flag settings, amounting to 31 conditional jumps in all (on the 8088/8086). Each of these jumps is a short jump of a maximum of 127 bytes in either direction. If a conditional jump has to be made outside this range, it has to be done by using a conditional jump to the position of an unconditional jump. Programs should therefore be planned so as to ensure that conditional jumps will as far as possible, occur within the 127 byte range.

Repetition

Loops can be written in assembly language using jumps to prior locations in the usual way, with other tests and jumps to implement ending a loop. In addition, the Intel set of processors provides the LOOP instruction which depends on the CX register to count out a specified number of iterations. The CX register is decremented on each iteration and the looping ends when the content of CX is zero. For example:

LOOP back

will return control to the position of label **back** for as long as CX contains a non-zero number, and control will pass to the following instruction when CX contains zero. The use of LOOP as an instruction can cause problems for programmers of other chips who are accustomed to using LOOP as a label name. Another snag is that if the CX register is not specifically loaded before the loop starts there is a danger that CX might contain zero at the start of the loop. This can be detected by using the JCXZ test instruction to by-pass the loop if CX is zero, or by testing CX and loading it if it contains zero.

The testing of the CX register that is carried out by LOOP does not involve the zero flag, so that there are two further versions of LOOP that can test the Z-flag as well as the CX content. Some assemblers use LOOPE and LOOPNE, others use LOOPZ, LOOPNZ, but the forms are equivalent. The LOOPZ/LOOPE instruction will branch if CX is not zero AND the zero flag is set. The LOOPNZ/LOOPNE instruction will branch if CX is not zero AND the zero flag is not set.

List and table actions

Loops in assembly language are often used in the processing of lists, particularly for string manipulation. Intel assembly language provides a set of string instructions which reduce the effort of writing loops for such purposes, and which use registers AX along with register pairs DS: SI (source) and ES: DI (destination). Each instruction can exist in byte or word form, designated by ending letters B or W, to specify whether the list element is a byte or a word. A special looping instruction, REP, can be used to force looping to continue until the number in CX has been decremented to zero.

For example, the statement:

LODSB

will load AX with the byte whose address is contained in DS: SI **and** the number in SI will be incremented by one byte after the load action. If LODSW had been used, a word would be loaded, and the SI value incremented by one **word** rather than one byte. Taking another example:

REP STOSB

would repeatedly store the AX content into the memory location provided by ES: DI, incrementing DI and decrementing CX until the CX content became zero.

The XLAT instruction is used to implement look-up tables. The base address of the table must be stored in BX, and the byte which is to be translated by the table is placed into AL. The action is to add the byte in AL to the address in BX and use the result as the address from which the translated byte is to be obtained. The byte is loaded into AL from that address. The BX address is not changed, and no flags are affected. XLAT is often used along with LODSB and STOSB and LOOP to perform translations of several bytes.

Subroutines/procedures

The concept of subroutines, familiar in 8-bit processors, is extended in the Intel 16 and 32-bit processors to allow for a **procedure** which might be contained in a separate file. Each procedure must be names, and will be taken as being **near** (in the same segment) unless **far** is specified. The return instruction for a near procedure will consist of two bytes; for a far procedure four bytes must be used. The procedure definition is headed by the word PROC and its end is marked by ENDP.

The procedure or subroutine is called using CALL in the usual way, and will end with some variety of RET. RET by itself specifies a return to the step following the CALL, and RET followed by a number will clear that number of locations from the stack. The A86 assembler offers in addition an alternative method which can dispense with PROC, treating the procedure as a simple routine with the near or far return specified in the RET instruction.

Parameters which are to be used by a procedure can be passed by value by loading numbers into a register before the CALL. Passing values by address can be implemented by loading a register with the address (offset) of a variable before calling the procedure. The other (slower) option is to pass variables or address offsets by using the stack.

The stack operates with word-sized units, permitting a word to be pushed from memory, from a register (including a segment register) or for more specialized purposes pushing the flags register. The POP operations are the reverse of the PUSH actions.

Ports and interrupts

The IN and OUT instructions deal with port transfers, in which the port number must be specified, either directly or as a number contained in the DX register. The syntax will be of the form:

IN AX, 05 ; input to AX from port 5

or

OUT DX, AL ; output from AL through port whose number is in DX

of which the latter method can make use of 16-bit port numbers, but the former can use only single-byte port addresses. Note the syntax of IN and OUT which follows the general format of destination, source.

The software interrupts are specified by using INT followed by the number of the interrupt, or using the form INTO (interrupt on overflow). The end of an interrupt routine is marked by IRET, interrupt return. Each INT will have the effect of pushing the flag registers, clearing the interrupt and trap flags and making a call to the address which is specified by a table entry corresponding to the number following INT. The interrupts 0 to 31 are reserved, and one of them, INT 4, is used by the INTO interrupt. The reserved interrupts have been listed in Section 10, and also the MS–DOS specific interrupts, notably INT 21H. Software interrupts can be disabled by clearing the interrupt-enable flag in the flags register, and the automatic action of clearing this flag at the start of an interrupt ensures that an interrupt routine cannot itself be interrupted unless it contains code for setting the interrupt flag. The trap flag causes the INT 1 interrupt to be generated after completing each instruction, so allowing single-stepping of instructions to be carried out by a debugging program.

```
Function 00H - used to terminate COM programs only. CS register must contain
program segment prefix segment address.
Function 01H - read key, echo on screen. Character code is returned in AL
register. Must be called twice to read function keys etc.
Function 02H - byte in DL register displayed on screen.
Function 03H - waits for serial port input, byte in AL register.
Function 04H - serial port output, byte in DL register.
Function 05H - parallel printer output, byte in DL register.
Function 06H - keyboard input or screen output depending on value in DL
register. If DL=FFH, then when a character is input the zero flag is cleared
and character code is in AL. If there is no input, the zero flag is set and 00H
is in AL. With DL=0 (or anything other than FFH), the code in DL is output to
the screen. Ctrl-Break is not obeyed.
Function 07H - key input, no echo, character in AL register, no action with
Ctrl-Break.
Function 08H - as 07H, but checks for Ctrl-Break.
Function 09H - print string whose address is in DS:DX. String end marked with
$ which is not printed
```

Figure 11.1 *The INT 21 functions.*
Note: the older types of 'control-block' file functions, and the most recent networking functions have been omitted. The function number is placed in the AH register and INT21H is called.

12 Systems design and typical usage

This brief final section consists mainly of circuit diagrams that illustrate typical systems usage of the various chips of the Intel 16/32 bit family, and these diagrams are reproduced here by courtesy of Intel and Elonex PLC. Details of larger systems are omitted – it would be pointless, for example, to show the precise wiring to each RAM chip in a computer system. Diagrams of the smaller systems show only the relevant pins connected.

Figure 12.1 shows the 8088 used in minimum configuration with a multiplexed bus. The MN/MX pin is connected to V_{cc} to enforce minimum system action, and connections are illustrated for two ports and the 8185 static RAM chip. The bus connects directly with the microprocessor, without buffers, and uses the multiplexed address/data lines of the 8088 directly. This is possible with the chips that are shown because the ports and the SRAM chips all use the ALE input to indicate when the combined address/data lines contain a valid address.

Figure 12.2 shows a demultiplexed bus configuration, still in minimum mode, for the 8088. The address and address/data lines are all latched, with the ALE output of the 8088 controlling the STB input of the latch chips. The system address bus is then available beyond the latch chips. The data bus for the system is obtained from a transceiver which is enabled by a signal from a gate that combines the EN output of the 8088 with the EN output from the interrupt controller chip, 8259A. No buffering is provided for the control signals in this layout.

Figure 12.3 shows the layout of a maximum-mode fully-buffered system, in which the local bus connects the 8088 to the 8288 bus controller, the latch and transceiver chips and the 8259A interrupt controller. These chips then provide the demultiplexed bus and the buffered control signals for the maximum-mode system.

The layouts of both maximum and minimum

mode systems for the 8086 follow along the same pattern as those for the 8088 with the obvious exception of the use of the AD0–AD15 bus on the 8086. The fully buffered and demultiplexed bus structures are as used in the older types of PC computer based on the 8088/8086 types of chips.

The 80188 and 80186 processors contain the equivalent of support chips internally, so that a system using these chips can be illustrated with reference to a computer rather than a simple system. Figure 12.4 shows a computer outline circuit for the 80186, using only the latch and transceiver chips between the 80186 and the specialized portions of the computer circuit. Note the use of low, middle and upper memory selection to separate the ROM from the program RAM (middle section) and data RAM (lower section).

The basic 80286 system is illustrated in Figure 12.5, showing the clock chip 82C284 used to supply timing and to synchronize RESET and READY. The system bus is obtained from the latch chips, the transceivers and the 8259A interrupt controller. This layout is very similar to that of the 8086 in its maximum mode. The addition of an 80287 co-processor was indicated earlier in Figure 6.19. Note in this diagram the elaborate decoding of the address bus which is required when these chips are used together.

When the 82258 DMA co-processor is added, the circuit of Figure 12.6 shows how this chip links into the local bus, and indicates the way in which the basic 20286 system can be extended.

The complete AT bus system is implemented for the 80286 or 80386SX type of chip by adding the 82230/82231 chips, as indicated in Figure 12.8, showing the five address and four data bus positions. The use of these support chips allows for the full 16M of memory to be used with the AT bus structure. The circuitry shows the 80286 used, but similar schemes can be used for the 80386DX and 80386SX processors.

IBM has chosen to use the MCA type of bus with the 80386DX and i486 processors, but other manufacturers have made use of the standard AT bus for which a large range of plug-in cards already exists. 80386DX machines can therefore be found using AT, MCA or EISA buses, causing considerable confusion among buyers.

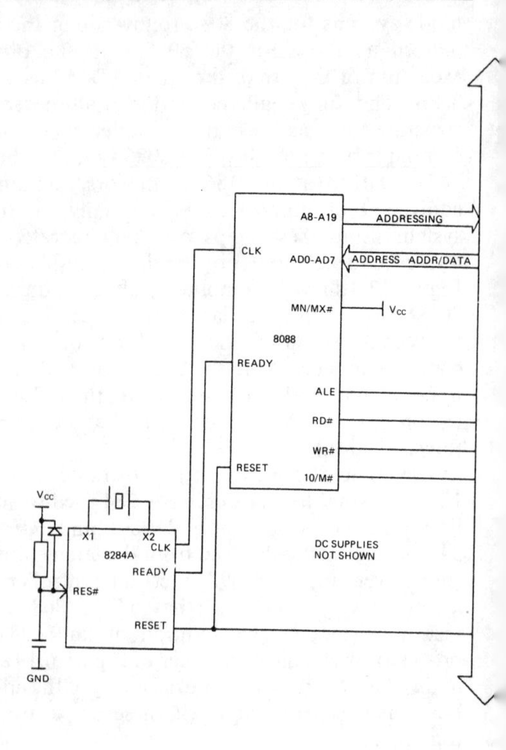

Figure 12.1 *A diagram for a minimum 8088 system using a multiplexed bus.*

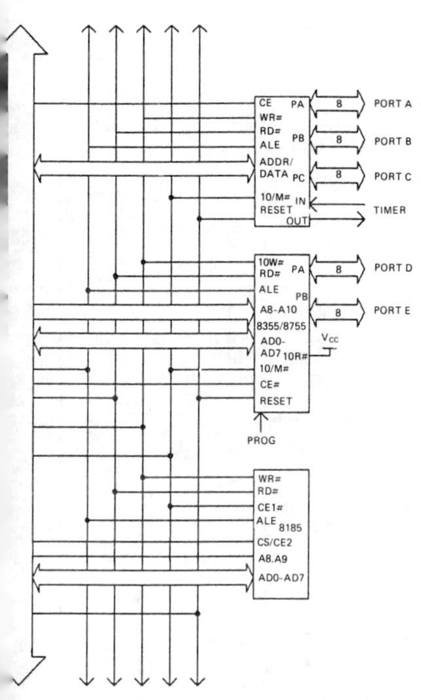
NTERNAL BUS LINES
CE
WR#
RD#
ALE
ADDR/
DATA
10/M#
RESET
PA
PB
PC
IN
OUT
8
PORT A
PORT B
PORT C
TIMER
10W#
RD#
ALE
A8-A10
8355/8755
AD0-
AD7
10/M#
CE#
RESET
10R#
Vcc
PORT D
PORT E
PROG
WR#
RD#
CE1#
ALE
8185
CS/CE2
A8.A9
AD0-AD7
TERNAL BUS LINES

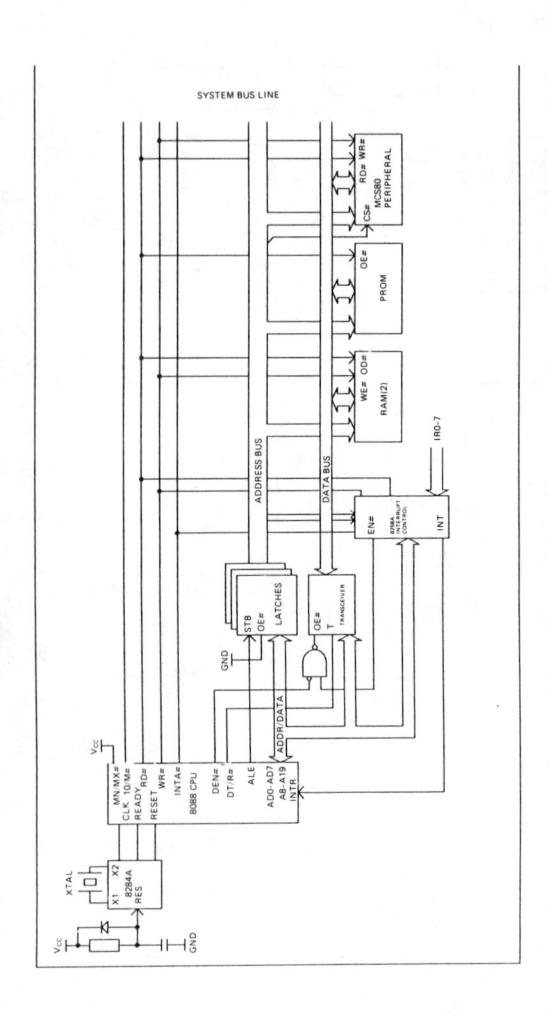

Figure 12.2 *An 8088 minimum system in which the buses have been demultiplexed to form a system bus set which can be more easily used by peripherals.*

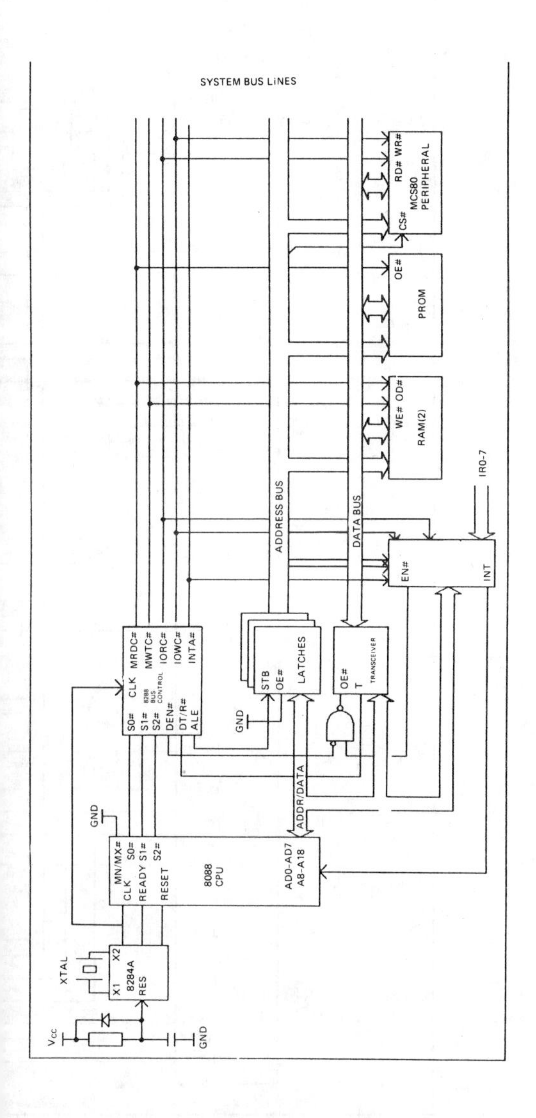

Figure 12.3 *The form of an 8088 maximum-mode system, in which the control bus is determined by the 8288 bus control chip. This is the type of format used for the original PC machines using the 8088.*

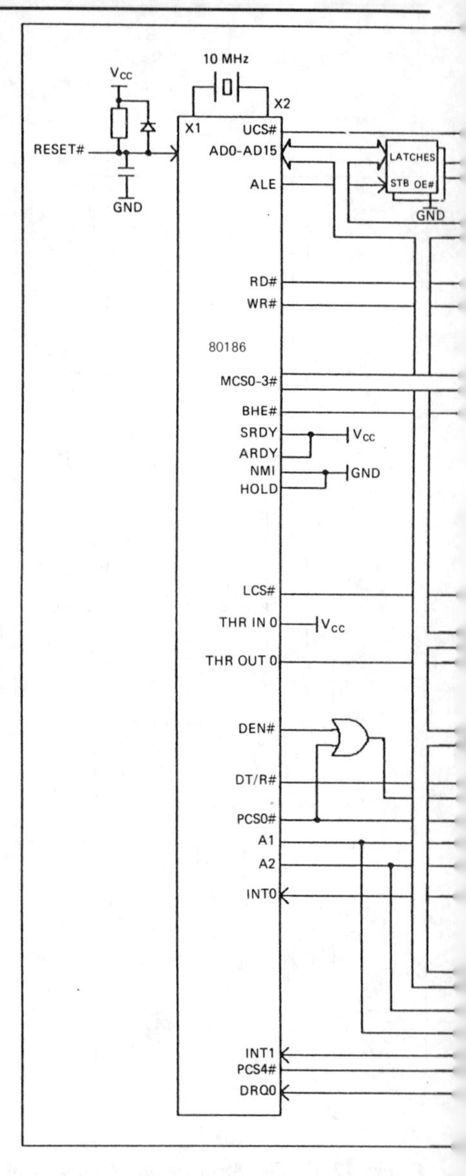

Figure 12.4 *Using the 80186 as the basis for a computer with very low chip-count.*

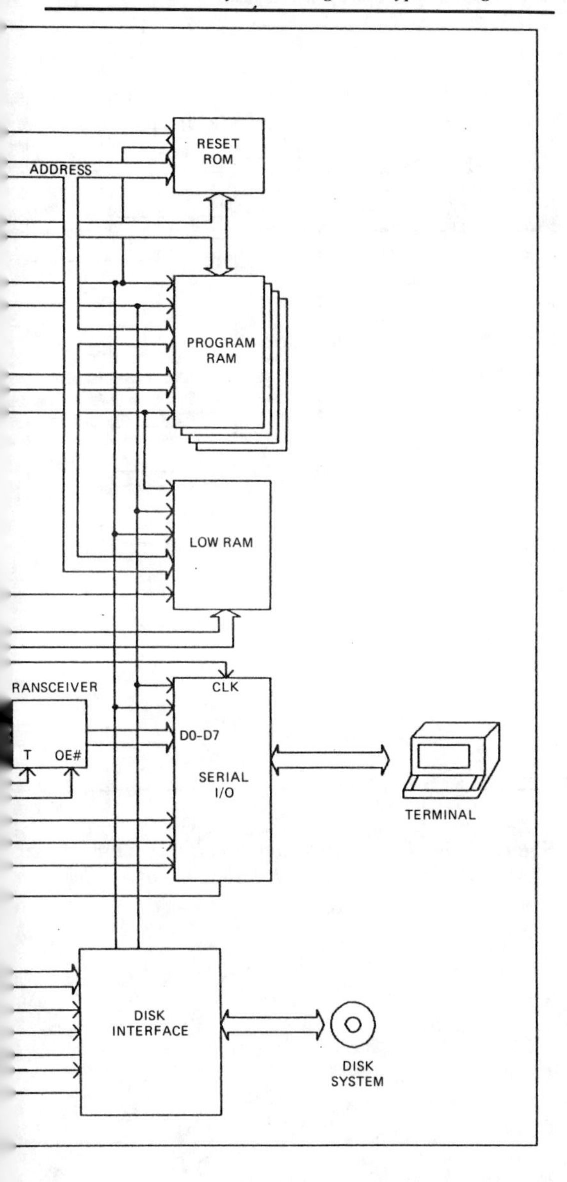
RESET
ROM
ADDRESS
PROGRAM
RAM
LOW RAM
RANSCEIVER
T
OE#
CLK
D0-D7
SERIAL
I/O
TERMINAL
DISK
INTERFACE
DISK
SYSTEM

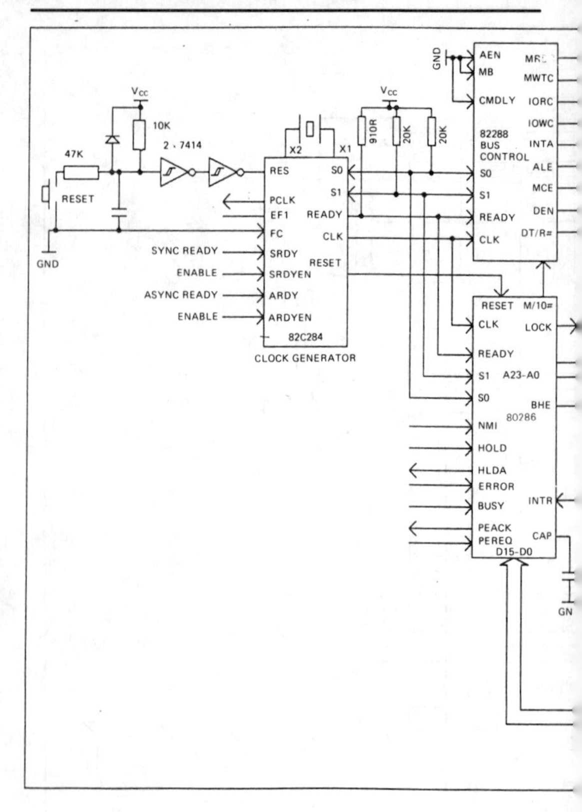

Figure 12.5 *A minimal 80286 circuit which uses the 82288 bus controller to establish the control bus lines.*

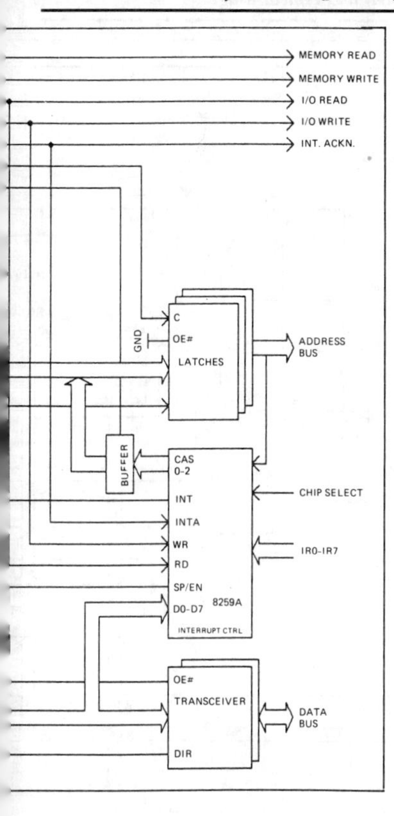
MEMORY READ
MEMORY WRITE
I/O READ
I/O WRITE
INT. ACKN.
C
OE#
GND
LATCHES
ADDRESS BUS
BUFFER
CAS 0-2
CHIP SELECT
INT
INTA
WR
IR0-IR7
RD
SP/EN
D0-D7
8259A
INTERRUPT CTRL
OE#
TRANSCEIVER
DATA BUS
DIR

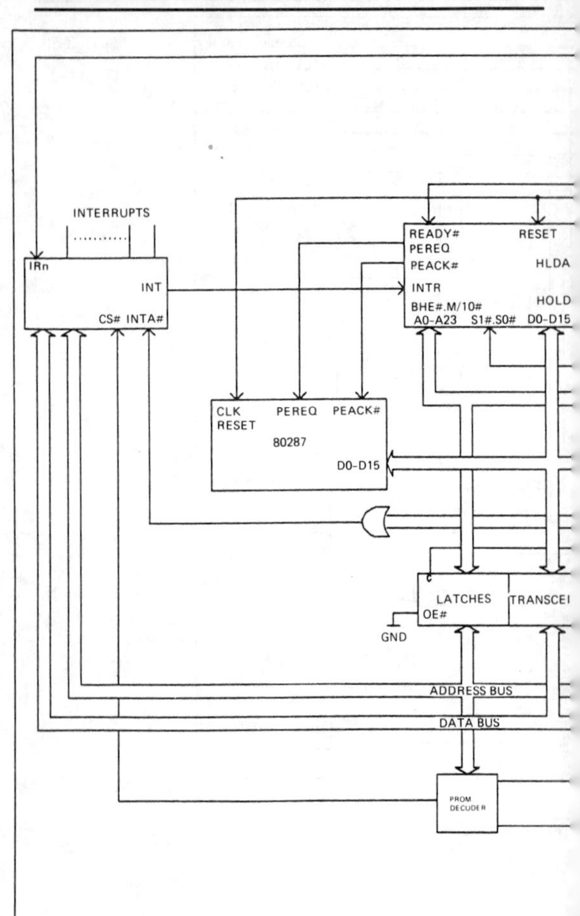

Figure 12.6 *A more elaborate 80286 circuit which uses the numerical co-processor and also the 82258 DMA co-processor.*

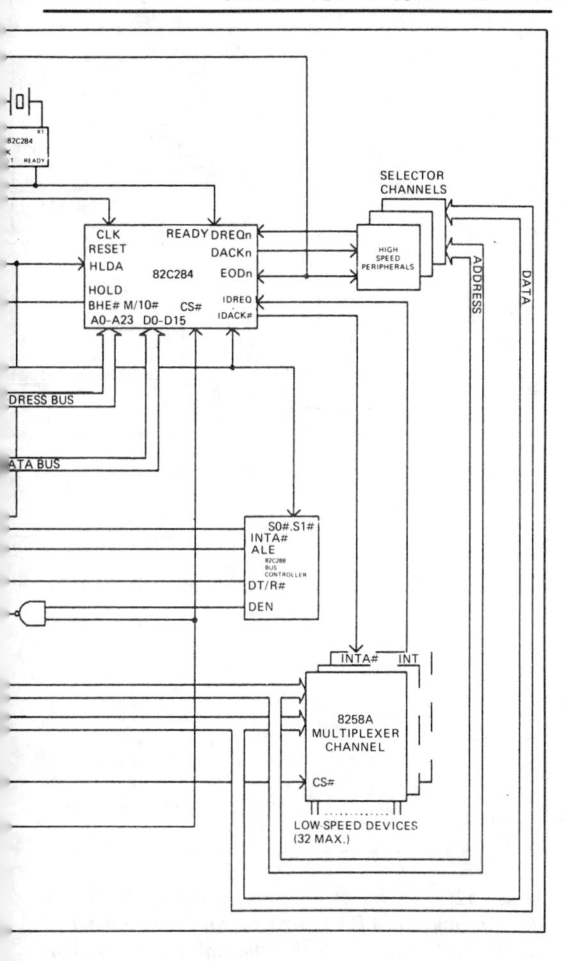
82C284
READY
SELECTOR
CHANNELS
CLK
RESET
HLDA
HOLD
BHE# M/10#
A0-A23 D0-D15
READY DREQn
DACKn
EODn
IDREQ
IDACK#
82C284
CS#
HIGH
SPEED
PERIPHERALS
ADDRESS
DATA
DRESS BUS
ATA BUS
S0#.S1#
INTA#
ALE
82C288
BUS
CONTROLLER
DT/R#
DEN
INTA#
INT
8258A
MULTIPLEXER
CHANNEL
CS#
LOW-SPEED DEVICES
(32 MAX.)

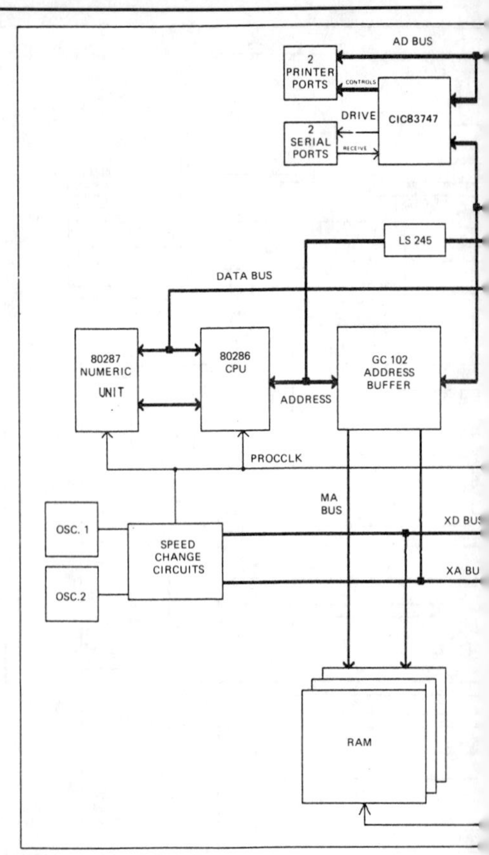

Figure 12.7 *A block diagram, courtesy of Elonex plc, showing the AT bus structure using a number of custom-built chips. This is a normal feature of computers using the AT bus.*

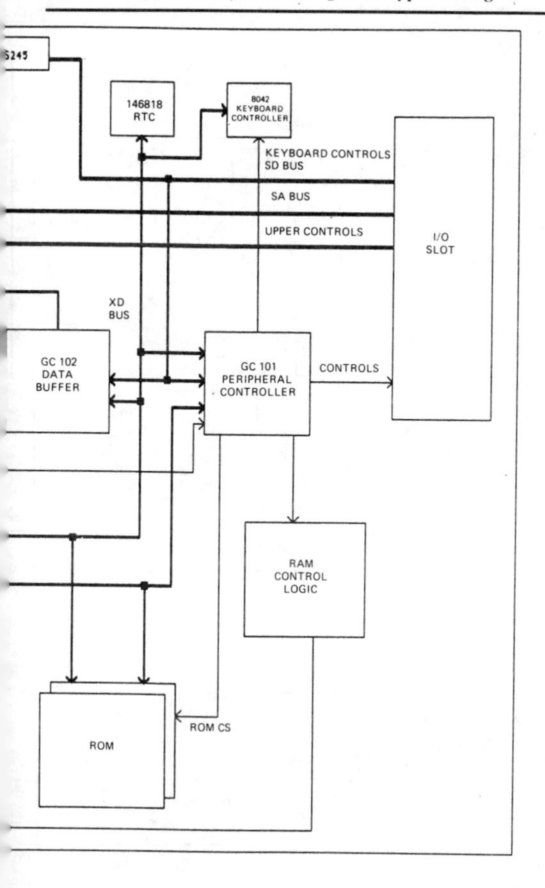

245
146818
RTC
8042
KEYBOARD
CONTROLLER
KEYBOARD CONTROLS
SD BUS
SA BUS
UPPER CONTROLS
I/O
SLOT
XD
BUS
GC 102
DATA
BUFFER
GC 101
PERIPHERAL
CONTROLLER
CONTROLS
RAM
CONTROL
LOGIC
ROM CS
ROM

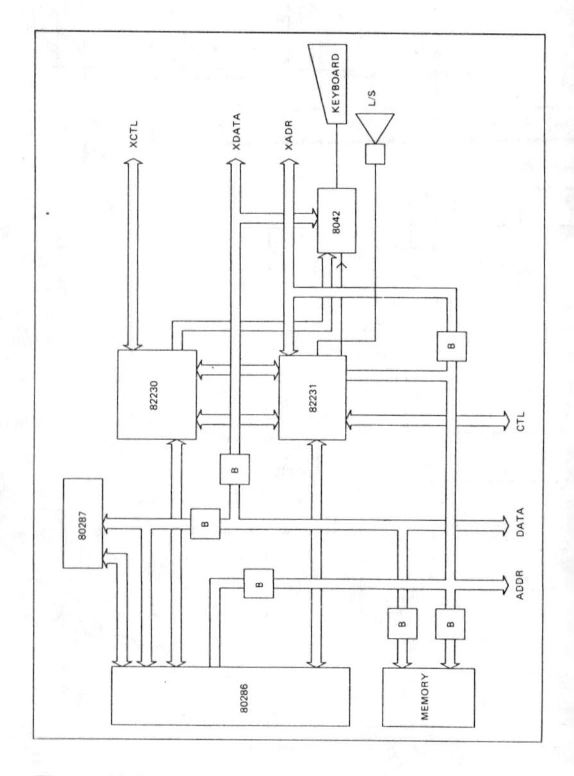

Figure 12.8 *An Intel circuit which shows the use of the 82230 and 82231 chips to establish the 'AT Bus' format for a 286 computer. Note the blocks marked 'B', which are buffers.*

The use of the AT bus along with the 80286 in a computer manufactured by Elonex is illustrated in Figure 12.7. This makes full use of a number of custom-designed chips which implement the AT bus standards, and is representative of the majority of computers which were being produced for the PC/AT market at the start of 1990.

Index